Lecture Notes in Mathematics

Volume 2285

This series reports on new developments in all areas of mathematics and their applications - quickly, informally and at a high level. Mathematical texts analysing new developments in modelling and numerical simulation are welcome. The type of material considered for publication includes:

1. Research monographs
2. Lectures on a new field or presentations of a new angle in a classical field
3. Summer schools and intensive courses on topics of current research.

Texts which are out of print but still in demand may also be considered if they fall within these categories. The timeliness of a manuscript is sometimes more important than its form, which may be preliminary or tentative.

Titles from this series are indexed by Scopus, Web of Science, Mathematical Reviews, and zbMATH.

More information about this series at http://www.springer.com/series/304

Atsushi Inoue

Tomita's Lectures on Observable Algebras in Hilbert Space

Springer

Atsushi Inoue
Department of Applied Mathematics
Fukuoka University
Fukuoka, Japan

ISSN 0075-8434 ISSN 1617-9692 (electronic)
Lecture Notes in Mathematics
ISBN 978-3-030-68892-9 ISBN 978-3-030-68893-6 (eBook)
https://doi.org/10.1007/978-3-030-68893-6

Mathematics Subject Classification: Primary: 46L99; Secondary: 46K10

This Springer imprint is published by the registered company Springer Nature Switzerland AG.
The registered company address is: Gewerbestrasse 11, 6330 Cham, Switzerland

Dedicated to the memory of my former supervisor late Professor Minoru Tomita

Preface

In 1967, M. Tomita presented his research on "standard forms of von Neumann algebras" at the international conference on C^*-algebras and their physical applications held at the University of Louisiana and at the fifth functional analysis symposium of Mathematical Society of Japan held in Sendai. This is an original theory that is the essence of noncommutative analysis. In 1970, M. Takesaki developed this theory and published the outcome of his work under the title "Tomita's Theory of Modular Hilbert Algebras and Its Applications" in Springer's Lecture Notes in Mathematics. This is known as the Tomita-Takesaki theory. I was a student of Professor Tomita from 1966 to 1971, and at that time, he did not mention to us anything about Tomita-Takesaki theory. He would come into the lecture room with a few sticks of chalks and lecture on research topics like "observable algebras", "operators and operator algebras on Krein spaces" and "noncommutative Fourier analysis" that were elaborated after the Tomita theory. At times, he was standing in front of the blackboard thinking and writing and suddenly he erased everything; it is certain that he was testing his mathematical thoughts of that moment. Personally, I could understand almost nothing of the contents of Tomita's lectures, so I was just keeping notes. However, I think that I naturally learned from him how to think about mathematics and how to approach mathematical problems. About 10 years later, I noticed that my research topic "unbounded operator algebras" is concerned with the theory of "observable algebras" presented in his aforementioned lectures. Thus, I started attending a lecture again for about 1 year. At that time, I got a copy of the previous lecture notes kept by my colleague H. Kurose, a student of Professor Tomita too, and read it myself in order to use in my research [16]. But, Tomita was not interested in publishing his results. So, as regards his research on "operators and operator algebras on Krein spaces", Y. Nakagami collaborated with him and the paper "Triangular Matrix Representation for Self-Adjoint Operators in Krein Spaces" was published in Japanese J. Math. in 1988 [23]. Nakagami continued his studies as the only author, resulting in the papers [21, 22]. Another of Tomita's students, S. Ôta, studied "Lorentz algebras on Krein spaces" [24]. The theory of observable algebras is closely related to operator algebras and its related fields. The Tomita-Takesaki theory is a special case of this theory, every observable algebra can be regarded as an operator algebra on a Pontryagin space with codimension 1 and the representation theory of locally convex $*$-algebras results in this theory. From all these, one concludes that this theory provides the mathematical techniques that establish intimate connections between the operator algebras and quantum theories. Unfortunately, Professor Tomita passed away in 2015 without publishing his work on observable algebras. Afraid that this theory would perish without being used, I decided to write these notes based on his research materials:

1. Notes of Professor Kurose and myself from Tomita's lectures.
2. Harmonic analysis on topological $*$-algebras [40].
3. Algebra of observables in Hilbert space [41].
4. Fundamental of noncommutative Fourier analysis [42].

There were many unproved parts, as well unclear parts in the preceding sources. This note is a compilation of the Tomita's observable algebras within the scope of what the author, who has been studying them for many years, could achieve.
First, I would like to thank my supervisor Professor M. Tomita for giving me many mathematical ideas and for his constant warm generous attention. Many thanks are also due to Professor Y. Nakagami for checking this manuscript in detail and for giving me much advice and comments. Furthermore, I would like to thank Professors M. Fragoulopoulou, S. Ôta and M. Uchiyama for many useful and helpful suggestions. I was able to complete this manuscript by giving lectures to Dr. H. Inoue and Dr. M. Takakura on the contents of the present work, in Fukuoka University, as well as having many discussions with them. I would like to thank H. Inoue for his careful reading of my manuscript and comments.

Fukuoka, Japan
July 2020

Atsushi Inoue

Contents

Chapter 1
Introduction

This note is a compilation of Tomita's observable algebras which the author has been studying for many years, based on the Tomita's lectures at Kyushu University and his RIM-workshop records [40, 41]. The theory of observable algebras is one of the mathematical techniques that establish intimate connections between the theory of operator algebras and quantum theories. In quantum mechanics, self-adjoint operators A in a Hilbert space $\mathcal{H}$ with inner product $(\cdot|\cdot)$ represent observables describing a given quantum system, while unit vectors x in $\mathcal{H}$ represent states of the system. The value $(Ax|x)$ for any state x is the expectation of the observable A on the state x. In the algebraic approach to quantum mechanics a $*$-algebra on $\mathcal{H}$ such as C^*-algebra or von Neumann algebra is often assigned to a given quantum mechanics system, and it has been studied from the mathematical aspects as well from the applications to quantum physics. The case that the correspondence of an operator observable A in a quantum system to a state x is a mapping has been considered often. If nothing like that happens in a physical phenomenon, how then can one approach this phenomenon mathematically? Tomita's idea for this question is to consider the quartet (A, x, y^*, μ) consisting of an observable A, of two states x, y and of an expectation μ as an observable, which is called a *quartet observable* on $\mathcal{H}$, where y^* is an element of the dual $\mathcal{H}^*$ of $\mathcal{H}$. This means that an operator observable A having two different states can be regarded as two different observables. From this approach of quartet observables, various types of observables are introduced and considered. The second idea is to introduce an algebraic and a topological structure into a set of quartet observables as follows: let $\mathcal{H}$ be a Hilbert space, $\mathcal{H}^*$ the dual of $\mathcal{H}$ and $B(\mathcal{H})$ the C^*-algebra of all bounded linear operators on $\mathcal{H}$. Let $Q^*(\mathcal{H})$ be the set of all quartet observables on $\mathcal{H}$. Define in $Q^*(\mathcal{H})$ the following operations $A + B$, αA, AB, the involution $A \to A^\sharp$ and the norm $\|A\|$ as follows: for $A = (A_0, x, y^*, \alpha)$, $B = (B_0, u, v^*, \beta) \in Q^*(\mathcal{H})$ and $\lambda \in \mathbb{C}$

$$\begin{aligned} A + B &= (A_0 + B_0, x + u, y^* + v^*, \alpha + \beta), \\ \lambda A &= (\lambda A_0, \lambda x, \lambda y^*, \lambda \alpha), \end{aligned}$$

A. Inoue, *Tomita's Lectures on Observable Algebras in Hilbert Space*,
Lecture Notes in Mathematics 2285, https://doi.org/10.1007/978-3-030-68893-6_1

$$AB = (A_0 B_0, A_o u, (B_0^* y)^*, (u|v)),$$
$$A^\sharp = (A_0^*, y, x^*, \bar{\alpha}),$$
$$\|A\| = \max(\|A_0\|, \|x\|, \|y\|, |\alpha|).$$

As seen above, a part of the dual state in a quartet observable is necessary for defining the algebraic structure (multiplication, involution) on $Q^*(\mathcal{H})$. From now on, we denote a quartet observable $A = (A_0, x, y^*, \alpha)$ by $A = (\pi(A), \lambda(A), \lambda^*(A), \mu(A))$. Then $Q^*(\mathcal{H})$ is a Banach $*$-algebra *without* identity, π is a bounded $*$-homomorphism of the Banach $*$-algebra $Q^*(\mathcal{H})$ onto the C^*-algebra $B(\mathcal{H})$, λ is a bounded linear mapping of $Q^*(\mathcal{H})$ onto $\mathcal{H}$ satisfying $\lambda(AB) = \pi(A)\lambda(B)$ for all $A, B \in Q^*(\mathcal{H})$, λ^* is a bounded linear mapping of $Q^*(\mathcal{H})$ onto $\mathcal{H}^*$ satisfying $\lambda^*(A) = \lambda(A^\sharp)^*$ and $\lambda^*(AB) = (\pi(B)^*\lambda(A^\sharp))^*$ for all $A, B \in Q^*(\mathcal{H})$ and μ is a bounded linear functional on $Q^*(\mathcal{H})$ satisfying $\mu(B^\sharp A) = (\lambda(A)|\lambda(B))$ for all $A, B \in Q^*(\mathcal{H})$. A $*$-subalgebra of $Q^*(\mathcal{H})$ is called a *Q^*-algebra* on $\mathcal{H}$ and a closed $*$-subalgebra of $Q^*(\mathcal{H})$ is called a *CQ^*-algebra* on $\mathcal{H}$. Tomita had explained the physical reason why he introduced such quartet observables. The start was a certain question to the Heisenberg and the von Neumann works about classical quantum mechanics. An observable means an observable part of a physical phenomenon. Heisenberg considered that an observable is expressed as a sequence (a_{ij}) of scalars. We call this a *sequential observation*. It is quite subjective since different observations must be possible. von Neumann considered that Heisenberg's sequential observation (a_{ij}) must be a representation of an operator A_0 on a Hilbert space $\mathcal{H}$: $(a_{ij}) = ((A_0 e_j|e_i))$ for some ONB $\{e_i\}$ in $\mathcal{H}$. Besides sequential observations, function observations are frequently used representing an observable to a certain space. He developed operator analysis in Hilbert space considering that an operator is an observable and an operator algebra is a dynamical system. But, it seems that the fundamental type of observables is an observable representation in our sense. For instance, the sequential observation (a_{ij}) of a Hilbert-Schmidt operator A_0 on $\mathcal{H}$ is different by taking the method of ONBs in $\mathcal{H}$. Because of this fact we consider the quartet observables. Indeed, we denote by $\mathcal{H} \otimes \bar{\mathcal{H}}$ the Hilbert space of all Hilbert-Schmidt operators on $\mathcal{H}$ with inner product defined by the trace of S^*T. For $A_0 \in \mathcal{H} \otimes \bar{\mathcal{H}}$ we consider the quartet observable $A = (\pi(A), \lambda(A), \lambda(A^\sharp)^*, \mu(A))$ on the Hilbert space $\mathcal{H} \otimes \bar{\mathcal{H}}$, where $\pi(A) = (a_{ij})$ (the Hilbert-Schmidt matrix of A_0), $\lambda(A) = A_0$, $\lambda(A^\sharp) = A_0^*$, $\mu(A) = \sum_{i=1}^{\infty} a_{ii}$. Then the map of the sequential observation (a_{ij}) to the quartet observable A is injective. We give a physical example of observables and observable algebras in Example 2.1.4. Tomita also introduced other observables and observable algebras as follows: for any $A_0 \in B(\mathcal{H})$ and $x, y \in \mathcal{H}$, the trio (A_0, x, y^*) is called a *trio observable*, and the set $T^*(\mathcal{H})$ of all trio observables on $\mathcal{H}$ is a Banach $*$-algebra without identity equipped with the following operations $A + B$, αA, AB, the involution $A \to A^\sharp$ and the norm $\|A\|$ defined as follows: for $A = (A_0, x, y^*)$, $B = (B_0, u, v^*)$ and $\alpha \in \mathbb{C}$

$$\begin{aligned} A + B &= (A_0 + B_0, x + u, y^* + v^*), \\ \lambda A &= (\lambda A_0, \lambda x, \lambda y^*), \\ AB &= (A_0 B_0, A_o u, (B_0^* y)^*), \\ A^\sharp &= (A_0^*, y, x^*), \\ \|A\| &= \max(\|A_0\|, \|x\|, \|y\|). \end{aligned}$$

A trio observable $A = (A_0, x, y^*)$ is denoted by $A = (\pi(A), \lambda(A), \lambda^*(A))$. Then π, λ and λ^* have the same properties as those in the case of $Q^*(\mathcal{H})$. A $*$-subalgebra of the Banach $*$-algebra $T^*(\mathcal{H})$ is called a *T^*-algebra* on $\mathcal{H}$ and a closed $*$-subalgebra of $T^*(\mathcal{H})$ is called a *CT^*-algebra* on $\mathcal{H}$. In Chap. 2 we show that every CQ^*-algebra is divided into two types by its expectation, and see that the study of CQ^*-algebras results in that of CT^*-algebras. So we mainly deal with T^*-algebras and CT^*-algebras. The main theme of this note is to investigate the relationship between the operator part $\pi(\mathfrak{A})$ and the vector part $\lambda(\mathfrak{A})$ of a CT^*-algebra $\mathfrak{A}$ on $\mathcal{H}$. The correspondence $\pi(A) \to \lambda(A)$ is not necessarily even a mapping. Using the fundamental theory of CT^*-algebras in Chaps. 2 and 3, we can prove that

(i) the correspondence $\pi(A) \to \lambda(A)$ is a uniformly continuous mapping from the $*$-algebra $\pi(\mathfrak{A})$ on $\mathcal{H}$ to the Hilbert space $\mathcal{H}$ if and only if there exists a vector g in $\mathcal{H}$ such that $\lambda(A) = \pi(A)g$ for all $A \in \mathfrak{A}$;
(ii) the correspondence $\pi(A) \to \lambda(A)$ is a uniformly closable mapping if and only if there exists a subset $\{g_\alpha\}$ of $\mathcal{H}$ such that $\lambda(A) = \sum_\alpha \pi(A) g_\alpha$ for all $A \in \mathfrak{A}$, where $\{g_\alpha\}$ is a subset of $\mathcal{H}$ satisfying $(\pi(A)g_\alpha | g_\beta) = 0$ for $\alpha \neq \beta$ and $\sum_\alpha \|\pi(A) g_\alpha\|^2 \leqq \|\lambda(A)\|^2$ for all $A \in \mathfrak{A}$;
(iii) $\lambda(N(\mathfrak{A}))$ is dense in $\mathcal{H}$ if and only if for any $x \in \mathcal{H}$ there exists a sequence $\{A_n\}$ in $\mathfrak{A}$ such that $\pi(A_n) = 0$ for all $n \in \mathbb{N}$ and $\lim_{n\to\infty} \lambda(A_n) = x$, where $N(\mathfrak{A}) := \{A \in \mathfrak{A};\ \pi(A) = 0\}$;
(iv) every CT^*-algebra is decomposed into a CT^*-algebra satisfying (ii) and a CT^*-algebra satisfying (iii).

In a physical sense, we regard a CT^*-algebra $\mathfrak{A}$ as a quantum mechanics system, its operator part $\pi(\mathfrak{A})$ as an operator observable algebra, its vector part $\lambda(\mathfrak{A})$ as a state space and its dual vector part $\lambda^*(\mathfrak{A})$ as a dual state space. It can be explained that a physical phenomenon is regular in case of (ii), and nonregular in case of (iii), and every physical phenomenon can be decomposed into the regular part and the nonregular part.

Chapters 2 and 3 build the foundations of the theory of T^*-algebras and CT^*-algebras. More precisely, Chap. 2 deals with a *spectrum* of a trio observable, a *functional calculus* of (self-adjoint) trio observable A, the *square root* of $A^\sharp A$ and the *polar decomposition* of A. After defining several locally convex topologies called *weak*, *σ-weak*, *strong*, *strong** and *σ-strong** on $T^*(\mathcal{H})$, we introduce three commutants $\mathfrak{A}^{\pi}$, $\mathfrak{A}^{\tau}$ and $\mathfrak{A}^{\rho}$ and the bicommutants $\mathfrak{A}^{\pi\pi}$, $\mathfrak{A}^{\tau\tau}$ and $\mathfrak{A}^{\rho\rho}$ of a T^*-algebra $\mathfrak{A}$. In Chap. 3 we treat the *von Neumann type density theorem* and the *Kaplansky type density theorem.* Both play an important role in the study of observable algebras.

Chapter 4 presents one of the major achievements of the theory, namely, the investigation of the relationship between the operator representation and the vector representation of a CT^*-algebra. Let $\mathfrak{A}$ be a CT^*-algebra on $\mathcal{H}$. We define the closed $*$-ideals $P(\mathfrak{A})$ and $N(\mathfrak{A})$ of $\mathfrak{A}$ by the closed linear span of the sets $\{AB;\ A, B \in \mathfrak{A}\}$ and $\{A \in \mathfrak{A};\ \pi(A) = 0\}$, respectively. Let $P_{\mathfrak{A}}$ and $N_{\mathfrak{A}}$ be the projections onto the closed subspace of $\mathcal{H}$ generated by the set $\pi(\mathfrak{A})\mathcal{H}$ and onto the closed subspace of $\mathcal{H}$ generated by the set $\lambda(N(\mathfrak{A}))$, respectively. Then $P_{\mathfrak{A}} \in \pi(\mathfrak{A})' \cap \pi(\mathfrak{A})''$ and $N_{\mathfrak{A}} \in \pi(\mathfrak{A})'$. Let $F_{\mathfrak{A}} = P_{\mathfrak{A}}(I - N_{\mathfrak{A}})$, and let $F(\mathfrak{A}) = \{(\pi(A), F_{\mathfrak{A}}\lambda(A), (F_{\mathfrak{A}}\lambda(A^{\sharp}))^*);\ A \in \mathfrak{A}\}$. Then we see that $F(\mathfrak{A})$ is a CT^*-algebra on $\mathcal{H}$. If $F_{\mathfrak{A}} = I$, $\mathfrak{A} = F(\mathfrak{A})$ and $\mathfrak{A} = N(\mathfrak{A})$, then $\mathfrak{A}$ is said to be regular, semisimple and singular, respectively, and it is proved that $\mathfrak{A}$ is decomposed into the semisimple CT^*-algebra $F(\mathfrak{A})$ and the singular CT^*-algebra $N(\mathfrak{A})$ as follows: $\mathfrak{A} = F(\mathfrak{A}) + N(\mathfrak{A})$, which shows that the aforementioned statement (iv) holds. Furthermore, proving the equalities: $P_{\mathfrak{A}} = I - N_{\mathfrak{A}^{\tau}}$ and $P_{\mathfrak{A}^{\tau}} = I - N_{\mathfrak{A}}$, we get one of the main results in this note: $\mathfrak{A}$ is semisimple if and only if the aforementioned statement (ii) holds, and $\mathfrak{A}$ is singular if and only if the aforementioned statement (iii) holds. This theory of observable algebras is concerned in that of operator algebras and its applications: in Chap. 2 we have seen that every observable algebra can be regarded as an operator algebra on a Pontryagin space of codimension 1. In Chap. 5 it is shown that the Tomita-Takesaki theory is a special case of this theory, the representation theory of (locally convex) $*$-algebras results in this theory, and this theory is applicable to the study of weights on C^*-algebras and of positive definite generalized functions in Lie groups. Since the author wrote this note on few research materials containing many unclear parts, there were among them some parts that he could not understand. So this note is a part of the work of Tomita within the range of author's understanding: therefore there are still many interesting subjects for investigation and thus there is room for further development. This note deals with an observable algebra whose operator representation is bounded, but the operator observables that appear in quantum mechanics are unbounded such as the position and momentum operators, and in addition the operator representation defined by the GNS-construction of a positive linear functional on a $*$-algebra is unbounded in general, From these, it is necessary to study an unbounded observable algebra, namely an observable algebra whose operator representation is unbounded. The ideas of Tomita have already influenced some of author's work for unbounded operator algebras such as the unbounded Tomita-Takesaki theory [3, 17] and the representation theory of (locally convex) $*$-algebras [3, 5, 16]. However, these studies are not contained in this note except for the author's study of admissible positive linear functionals on (locally convex) $*$-algebras contained in Sect. 5.1. The author's motivation was to write this note with the aim that Tomita's theory be widely known and freely used by researchers in many (unbounded) operator algebras and related fields without being buried. The systematic studies of unbounded observable algebras will be an issue for the future.

Chapter 2
Fundamentals of Observable Algebras

This chapter is devoted to the fundamental theory of observable algebras. In Sect. 2.1 the notions of observables and observable algebras are defined. Section 2.2 deals with structure of CQ^*-algebras. In Sect. 2.3 we discuss the spectrum of a trio observable and a functional calculus of a (self-adjoint) trio observable, and using it we define the square root and the polar decomposition of a trio observable. Section 2.4 discuss $*$-automorphisms of observable algebras. In Sect. 2.5 we define some locally convex topologies on $T^*(\mathcal{H})$ called weak, σ-weak, strong, σ-strong, strong* and σ-strong* and investigate their properties. In Sect. 2.6 we define and examine three commutants $\mathfrak{A}^\pi$, $\mathfrak{A}^\tau$ and $\mathfrak{A}^\rho$, and bicommutants $\mathfrak{A}^{\pi\pi}$, $\mathfrak{A}^{\tau\tau}$ and $\mathfrak{A}^{\rho\rho}$ of a T^*-algebra $\mathfrak{A}$.

2.1 Q^*-algebras and T^*-algebras

Let $\mathcal{H}$ be a Hilbert space with inner product $(\cdot|\cdot)$ and $\mathcal{H}^*$ the dual of $\mathcal{H}$. For any $x \in \mathcal{H}$ we define a linear functional x^* on $\mathcal{H}$ by

$$< x^*, y >= (y|x), \quad y \in \mathcal{H}.$$

Then $x^* \in \mathcal{H}^*$, and by the Riesz theorem the map: $x \to x^*$ is a conjugate linear isometry of $\mathcal{H}$ onto $\mathcal{H}^*$. Let $B(\mathcal{H})$ be the C^*-algebra of all bounded linear operators on $\mathcal{H}$. For $A_0 \in B(\mathcal{H})$ and $x \in \mathcal{H}$, we define elements x^*A_0 and A_0x^* of $\mathcal{H}^*$ by $x^*A_0 = (A_0^*x)^*$ and $A_0x^* = (A_0x)^*$. A quartet (A_0, x, y^*, r) of $A_0 \in B(\mathcal{H})$, $x, y \in \mathcal{H}$ and $r \in \mathbb{C}$ is called a *quartet observable* on $\mathcal{H}$.

A. Inoue, *Tomita's Lectures on Observable Algebras in Hilbert Space*,
Lecture Notes in Mathematics 2285, https://doi.org/10.1007/978-3-030-68893-6_2

Let $Q^*(\mathcal{H})$ be the set of all quartet observables on $\mathcal{H}$; this is a Banach $*$-algebra without identity equipped with the following operations, involution $\sharp$ and norm:

$$\begin{aligned} A+B &= (A_0+B_0, x+u, y^*+v^*, r+\delta), \\ \alpha A &= (\alpha A_0, \alpha x, \alpha y^*, \alpha r), \\ AB &= (A_0B_0, A_0u, (B_0^*y)^*, (u|y)), \\ A^\sharp &= (A_0^*, y, x^*, \bar{r}), \\ \|A\| &= \max(\|A_0\|, \|x\|, \|y\|, |r|) \end{aligned}$$

for $A = (A_0, x, y^*, r)$, $B = (B_0, u, v^*, \delta) \in Q^*(\mathcal{H})$ and $\alpha \in \mathbb{C}$. For $A = (A_0, x, y^*, r) \in Q^*(\mathcal{H})$, write

$$\pi(A) = A_0, \quad \lambda(A) = x, \quad \lambda^*(A) = y^*, \quad \mu(A) = r.$$

Then π is a bounded $*$-homomorphism of the Banach $*$-algebra $Q^*(\mathcal{H})$ onto the C^*-algebra $B(\mathcal{H})$ satisfying

$$\|\pi(A)\| \leq \|A\|$$

for all $A \in Q^*(\mathcal{H})$, λ is a bounded linear map of $Q^*(\mathcal{H})$ onto $\mathcal{H}$ satisfying

$$\lambda(AB) = \pi(A)\lambda(B)$$

for all $A, B \in Q^*(\mathcal{H})$, λ^* is a bounded linear map of $Q^*(\mathcal{H})$ onto $\mathcal{H}^*$ satisfying

$$\begin{aligned} \lambda^*(A) &= \lambda(A^\sharp)^*, \\ \lambda^*(AB) &= (\pi(B)^*\lambda(A^\sharp))^* \end{aligned}$$

for all $A, B \in Q^*(\mathcal{H})$, and μ is a bounded linear functional on $Q^*(\mathcal{H})$ satisfying

$$\mu(B^\sharp A) = (\lambda(A)|\lambda(B))$$

for all $A, B \in Q^*(\mathcal{H})$. We call π, λ, λ^* and μ the operator representation, the vector representation into $\mathcal{H}$, the vector representation into $\mathcal{H}^*$ and the expectation of $Q^*(\mathcal{H})$, respectively.

Definition 2.1.1 A $*$-subalgebra of the Banach $*$-algebra $Q^*(\mathcal{H})$ is called a quartet observable algebra on $\mathcal{H}$, for simplicity, a Q^*-algebra on $\mathcal{H}$. A closed $*$-subalgebra of the Banach $*$-algebra $Q^*(\mathcal{H})$ is called a complete quartet observable algebra on $\mathcal{H}$, for simplicity, a CQ^*-algebra on $\mathcal{H}$.

We next define the notions of trio observable algebras. A triple (A_0, x, y^*) of $A_0 \in B(\mathcal{H})$, $x \in \mathcal{H}$ and $y^* \in \mathcal{H}^*$ is called a *trio observable* on $\mathcal{H}$, and denote by $T^*(\mathcal{H})$ the set of all trio observables on $\mathcal{H}$. Then $T^*(\mathcal{H})$ is a Banach $*$-algebra without identity equipped with the following operations, involution and norm:

$$\begin{aligned} A + B &= (A_0 + B_0, x + u, y^* + v^*), \\ \alpha A &= (\alpha A_0, \alpha x, \alpha y^*), \\ AB &= (A_0 B_0, A_0 u, (B_0^* y)^*), \\ A^\sharp &= (A_0^*, y, x^*), \\ \|A\| &= \max(\|A_0\|, \|x\|, \|y\|) \end{aligned}$$

for $A = (A_0,\ x,\ y^*)$, $B = (B_0,\ u,\ v^*)$ and $\alpha \in \mathbb{C}$. For $A = (A_0,\ x,\ y^*) \in T^*(\mathcal{H})$, write

$$\pi(A) = A_0, \quad \lambda(A) = x \quad \text{and} \quad \lambda^*(A) = y^*.$$

Then π is a bounded $*$-homomorphism of the Banach $*$-algebra $T^*(\mathcal{H})$ onto $B(\mathcal{H})$, λ is a bounded linear map of $T^*(\mathcal{H})$ onto $\mathcal{H}$ satisfying

$$\lambda(AB) = \pi(A)\lambda(B)$$

for all $A, B \in T^*(\mathcal{H})$, and λ^* is a bounded linear map of $T^*(\mathcal{H})$ onto $\mathcal{H}^*$ satisfying

$$\begin{aligned} \lambda^*(A) &= \lambda(A^\sharp)^*, \\ \lambda^*(AB) &= (\pi(B)^*\lambda(A^\sharp))^* \end{aligned}$$

for all $A, B \in T^*(\mathcal{H})$.

Definition 2.1.2 A $*$-subalgebra of the Banach $*$-algebra $T^*(\mathcal{H})$ is called a trio observable algebra on $\mathcal{H}$, for simplicity, a T^*-algebra on $\mathcal{H}$. A closed $*$-subalgebra of the Banach $*$-algebra $T^*(\mathcal{H})$ is called a complete trio observable algebra on $\mathcal{H}$, for simplicity, a CT^*-algebra on $\mathcal{H}$.

Let $\mathfrak{A}$ be a T^*-algebra (or a Q^*-algebra) on $\mathcal{H}$. The norm topology on the Banach $*$-algebra $\mathfrak{A}$ is called the *uniform topology* on $\mathfrak{A}$ and is denoted by τ_u. Let $\mathfrak{A}$ and $\mathfrak{B}$ be T^*-algebras (or Q^*-algebras) on $\mathcal{H}$. If there exists an algebraic $*$-isomorphism and isometry of $\mathfrak{A}[\tau_u]$ onto $\mathfrak{B}[\tau_u]$, then we say that the T^*-algebras (or Q^*-algebras) $\mathfrak{A}$ and $\mathfrak{B}$ are *isomorphic*.

We investigate the relationship between $Q^*(\mathcal{H})$ and $T^*(\mathcal{H})$. An element $V := (0, 0, 0, 1)$ of $Q^*(\mathcal{H})$ is called the *vacuum* (or, *unit expectation*) of $Q^*(\mathcal{H})$, and $\mathbb{C}V := \{\alpha V;\ \alpha \in \mathbb{C}\}$ is identical with the radical of $Q^*(\mathcal{H})$, that is,

$$\mathbb{C}V = \{A \in Q^*(\mathcal{H});\ AX = XA \ \text{ for all } \ X \in Q^*(\mathcal{H})\}.$$

Hence $\mathbb{C}V$ is a closed $*$-ideal of $Q^*(\mathcal{H})$. Put

$$\tau(A) = (\pi(A),\ \lambda(A), \lambda^*(A)), \quad A \in Q^*(\mathcal{H}).$$

Then τ is a bounded $*$-homomorphism of $Q^*(\mathcal{H})$ onto $T^*(\mathcal{H})$ whose kernel is $\mathbb{C}V$. Hence $T^*(\mathcal{H})$ is isomorphic to the quotient Banach $*$-algebra $Q^*(\mathcal{H})/_{\mathbb{C}V}$ as Banach $*$-algebras.

Here we give two concrete examples of observable algebras. The first is of observable algebras whose elements are certain functions on $\mathbb{R}$.

Example 2.1.3 We denote by $C(\mathbb{R})$, $C_b(\mathbb{R})$, $L^1(\mathbb{R})$ and $L^2(\mathbb{R})$ the space of all continuous, bounded continuous, integrable and square integrable functions on $\mathbb{R}$, respectively.

(1) We put

$$\mathfrak{A}_t = \{f_t = (\pi(f), \lambda(f), \lambda(\bar{f})^*);\ f \in C_b(\mathbb{R}) \cap L^2(\mathbb{R})\},$$
$$\mathfrak{A}_q = \{f_q = (\pi(f), \lambda(f), \lambda(\bar{f})^*, \mu(f));\ f \in C_b(\mathbb{R}) \cap L^1(\mathbb{R})\},$$

where $\pi(f)g = fg$, $g \in L^2(\mathbb{R})$, $\lambda(f) = f$, $\lambda^*(f) = \bar{f}^*$ $(\bar{f}(x) = \overline{f(x)})$ and $\mu(f) = \int_{\mathbb{R}} f(x)dx$. Then $\mathfrak{A}_t$ is a CT^*-algebra on $L^2(\mathbb{R})$ which is isomorphic to the Banach $*$-algebra $C_b(\mathbb{R}) \cap L^2(\mathbb{R})$ with the usual function operations, involution and the norm $\|f\|_{u\wedge 2} := \max(\|f\|_u, \|f\|_2)$, where $\|f\|_u = \max_{t\in\mathbb{R}} |f(t)|$. $\mathfrak{A}_q$ is a CQ^*-algebra on $L^2(\mathbb{R})$ which is isomorphic to the Banach $*$-algebra $C_b(\mathbb{R}) \cap L^1(\mathbb{R})$ with the usual function operations, involution and the norm $\|f\|_{u\wedge 1} := \max(\|f\|_u, \|f\|_1)$.

(2) We put

$$\mathfrak{B}_t = \{f_t = (\pi(f), \lambda(f), \lambda(\bar{f})^*);\ f \in L^1(\mathbb{R}) \cap L^2(\mathbb{R})\},$$
$$\mathfrak{B}_q = \{f_q = (\pi(f), \lambda(f), \lambda(\bar{f})^*, f(0));\ f \in L^1(\mathbb{R}) \cap L^2(\mathbb{R}) \cap C(\mathbb{R})\},$$

where $\pi(f)q = f*q$ (the convolution operator, $(f*g)(x) = \int f(t)g(t-x)dt$), $\lambda(f) = f$, $\lambda^*(f) = \overline{f(-x)}^*$ and $\mu(f) = f(0)$. Then $\mathfrak{B}_t$ is a CT^*-algebra on $L^2(\mathbb{R})$ which is isomorphic to the Banach $*$-algebra $L^1(\mathbb{R}) \cap L^2(\mathbb{R})$ with the convolution multiplication $f*g$, the involution $\bar{f}(-x)$ and the norm $\|f\|_{1\wedge 2} := \max(\|f\|_1, \|f\|_2)$. $\mathfrak{B}_q$ is a CQ^*-algebra on $L^2(\mathbb{R})$ which is isomorphic to the Banach $*$-algebra with same multiplication and involution as those of $\mathfrak{B}_t$ and the norm $\|f\|_{u\wedge 1\wedge 2} := \max(\|f\|_u, \|f\|_1, \|f\|_2)$.

The second is of observable algebras related to the CCR-algebra of degree 1.

Example 2.1.4 Let $\mathscr{A}$ be a $*$-algebra generated by identity $\mathbb{1}$ and two Hermitian elements p and q satisfying the Heisenberg commutation relation:

$$[p, q] := pq - qp = -i\,\mathbb{1} \tag{2.1.1}$$

and it is called a CCR-algebra of degree 1. It is not possible to find Hermitian matrices P and Q satisfying the Heisenberg commutation, however Heisenberg found a solution in the form of infinite matrices:

$$P = \frac{-i}{\sqrt{2}} \begin{pmatrix} 0 & 1 & 0 & 0 & 0 & \cdots \\ -1 & 0 & \sqrt{2} & 0 & 0 & \cdots \\ 0 & -\sqrt{2} & 0 & \sqrt{3} & 0 & \cdots \\ 0 & 0 & -\sqrt{3} & 0 & \sqrt{4} & \cdots \\ 0 & 0 & 0 & -\sqrt{4} & 0 & \cdots \end{pmatrix},$$

$$Q = \frac{1}{\sqrt{2}} \begin{pmatrix} 0 & 1 & 0 & 0 & \cdots \\ 1 & 0 & \sqrt{2} & 0 & \cdots \\ 0 & \sqrt{2} & 0 & \sqrt{3} & \cdots \\ 0 & 0 & \sqrt{3} & 0 & \cdots \end{pmatrix}. \tag{2.1.2}$$

Schrödinger found another solution to the Heisenberg solution in his formulation of quantum mechanics. The operators P_S and Q_S defined by

$$(P_S f)(t) = -i\frac{d}{dt}f(t),$$
$$(Q_S f)(t) = tf(t)$$

for every function f in the Schwartz space $S(\mathbb{R})$ are essentially self-adjoint operators in $L^2(\mathbb{R})$ satisfying the Heisenberg commutation relation. We here denote by the same P_S and Q_S the closures of P_S and Q_S, which are called the momentum and position operators, respectively. He showed that P_S and Q_S have the same matrices as in Eq. (2.1.2) with respect to the ONB for $L^2(\mathbb{R})$ defined by

$$\varphi_n(t) = \pi^{-\frac{1}{4}}(2^n n!)^{-\frac{1}{2}}(t - \frac{d}{dt})^n e^{-\frac{t^2}{2}}.$$

This representation is called the Schrödinger representation of the CCR-algebra $\mathscr{A}$. This raised the question as to whether there were other ways to realize the Heisenberg commutation relations. Weyl proposed the question should be formulated in terms of strongly continuous unitary groups $U(s) = e^{isP}$ and $V(t) = e^{itQ}$ satisfying the commutation relation:

$$U(s)V(t) = e^{ist}V(t)U(s) \tag{2.1.3}$$

for all $s, t \in \mathbb{R}$. Von Neumann solved the question posed by Weyl. He showed that

$$(U_S(t)f)(x) := (e^{itP_S}f)(x) = f(x+t),$$
$$(V_S(t)f)(x) := (e^{itQ_S}f)(x) = e^{itx}f(x)$$

for all $f \in L^2(\mathbb{R})$ and $t \in \mathbb{R}$ and $\{U_S(s), V_S(t);\ s, t \in \mathbb{R}\}$ satisfies the Weyl commutation relation (2.1.3), and every strong continuous unitary groups $\{U(s);\ s \in \mathbb{R}\}$ and $\{V(t);\ t \in \mathbb{R}\}$ on $\mathcal{H}$ satisfying the Weyl commutation relation is unitarily equivalent to a direct sum of a sequence of the unitary groups $\{U_S(s);\ s \in \mathbb{R}\}$ and $\{V_S(t);\ t \in \mathbb{R}\}$. The unitary groups $\{U_S(s);\ s \in \mathbb{R}\}$ and $\{V_S(t);\ t \in \mathbb{R}\}$ defined by the momentum operator P_S and the position operator Q_S are called motion groups. Here we remark that the unitary groups $\{e^{isP}, e^{itQ};\ s, t \in \mathbb{R}\}$ defined by self-adjoint operators P and Q satisfying the Heisenberg commutation relation don't necessarily satisfy the Weyl commutation relation. The momentum operator P_S and the position operator Q_S themselves must be observables , as well various observables whose operator parts are defined by P_S and Q_S. Thus there are observables whose operator parts are unbounded. From this, we here after need to consider such observables and observable algebras. This note deals with only observables whose operator parts are bounded and observable algebras consisting of such observables. Hence we consider observable algebras whose operator parts are contained in a $*$-algebra defined by the unitary groups $\{U_S(s);\ s \in \mathbb{R}\}$ and $\{V_S(t);\ t \in \mathbb{R}\}$. Then we can introduce various types of observables and observable algebras. For instance, let $f_0 \in L^2(\mathbb{R})$. Then $f_t := U_S(t)f_o,\ t \in \mathbb{R}$ must be a motive observable, expressing the time process of vectors. This processing must be the solution of motion equation $\frac{d}{dt}x(t) = iP_S x(t)$. Then we can think that the momentum operator P_S is an operator observable, which determines the process of f_t by knowing the initial value f_0. We can discuss the same thing for the position operator Q_S. Let $\mathfrak{A}_0$ be a $*$-algebra on $L^2(\mathbb{R})$ defined by $\{U_S(s);\ s \in \mathbb{R}\}$ and $\{V_S(t);\ t \in \mathbb{R}\}$. For any non-zero element f in $L^2(\mathbb{R})$ we put

$$\mathfrak{A}_f := \{A = (A_0, A_0 f, (A_0^* f)^*, (A_0 f | f));\ A_0 \in \mathfrak{A}_0\}.$$

Then it is a Q^*-algebra whose operator representation, vector and dual representations and expectation are well-behaved. On the other hand, we put

$$\mathfrak{B} := \{(0, f, g^*, \alpha);\ f, g \in L^2(\mathbb{R}),\ \alpha \in \mathbb{C}\}.$$

Then it is a Q^*-algebra on $L^2(\mathbb{R})$ whose operator, vector, dual vector representations and expectation don't have any relationships.

Finally we show that $Q^*(\mathcal{H})$ can be regarded as an operator algebra on the Krein space $\mathbb{C}\times\mathcal{H}\times\mathbb{C}$. We first briefly recall some definitions and terminology concerning Krein spaces. Suppose that J is an Hermitian unitary operator on a Hilbert space $\mathcal{H}$ with inner product $(\cdot|\cdot)$. Then the Hermitian sesquilinear form $< \cdot, \cdot >_J$ on $\mathcal{H} \times \mathcal{H}$ defined by

$$< x, y >_J = (Jx|y), \quad x, y \in \mathcal{H},$$

is called an *indefinite inner product* on $\mathcal{H}$ defined by J, and $\mathcal{H}$ is said to be a *Krein space* with the indefinite inner product $< \cdot, \cdot >_J$, and is denoted by $\mathcal{H}_J$. By the

Riesz theorem, for any $A \in B(\mathcal{H})$ there exists a unique element A^J of $B(\mathcal{H})$ such that

$$< Ax, y >_J = < x, A^J y >_J$$

for all $x,\ y \in \mathcal{H}$, which satisfies the equality:

$$A^J = JA^*J.$$

The operator A^J is called the J*-adjoint* of A. By the spectral resolution of J, there exist projections J_+ and J_- satisfying

$$J = J_+ - J_-.$$

The closed subspaces $J_+\mathcal{H}$ and $J_-\mathcal{H}$ of $\mathcal{H}$ are denoted by $\mathcal{H}_+$ and $\mathcal{H}_-$, and their dimensions are denoted by dim $\mathcal{H}_+$ and dim $\mathcal{H}_-$, respectively. If dim $\mathcal{H}_+$ and dim $\mathcal{H}_-$ are both finite, then the Krein space $\mathcal{H}_J$ is called the *Minkowski space*, and if either dim $\mathcal{H}_+$ or $\mathcal{H}_-$ is finite, then $\mathcal{H}_J$ is called the *Pontrjagin space*. We now consider the product space $\mathcal{K} := \mathbb{C} \times \mathcal{H} \times \mathbb{C}$, and write an element x of $\mathcal{K}$ by

$$x = \begin{pmatrix} \mu(x) \\ \lambda(x) \\ \delta(x) \end{pmatrix},$$

where $\mu(x),\ \delta(x) \in \mathbb{C}$ and $\lambda(x) \in \mathcal{H}$. Then $\mathcal{K}$ is a Hilbert space under the usual operations and inner product:

$$(x|y) = \mu(x)\overline{\mu(y)} + (\lambda(x)|\lambda(y)) + \delta(x)\overline{\delta(y)}$$

for all $x,\ y \in \mathcal{K}$, and has an Hermitian unitary operator J on $\mathcal{K}$:

$$J = \begin{pmatrix} 0 & 0 & 1 \\ 0 & I & 0 \\ 1 & 0 & 0 \end{pmatrix}.$$

Hence, $\mathcal{K}$ is a Krein space with indefinite inner product:

$$< x, y >_J = (Jx|y) = \mu(x)\overline{\delta(y)} + (\lambda(x)|\lambda(y)) + \delta(x)\overline{\mu(y)}$$

for all $x,\ y \in \mathcal{K}$. Since

$$J_+ = \begin{pmatrix} 1/2 & 0 & 1/2 \\ 0 & I & 0 \\ 1/2 & 0 & 1/2 \end{pmatrix} \quad \text{and} \quad J_- = \begin{pmatrix} 1/2 & 0 & -1/2 \\ 0 & 0 & 0 \\ -1/2 & 0 & 1/2 \end{pmatrix},$$

it follows that $\dim \mathcal{K}_+ = \infty$ and $\dim \mathcal{K}_- = 1$. Thus the Krein space $\mathcal{K}_J$ is a Pontrjagin space with codimension 1. An element A of $Q^*(\mathcal{H})$ is represented as an operator on the Pontrjagin space $\mathcal{K}_J$:

$$A = \begin{pmatrix} 0 & \lambda^*(A) & \mu(A) \\ 0 & \pi(A) & \lambda(A) \\ 0 & 0 & 0 \end{pmatrix}.$$

The involution $A^\sharp$ of $A \in Q^*(\mathcal{H})$ coincides with the J-adjoint A^J:

$$A^J = \begin{pmatrix} 0 & \lambda(A)^* & \overline{\mu(A)} \\ 0 & \pi(A)^* & \lambda^*(A)^* \\ 0 & 0 & 0 \end{pmatrix}.$$

Thus Q^*-algebras can be regarded as operator algebras on Pontrjagin spaces with codimension 1.

T^*-algebras are also related to operator algebras on Pontrjagin spaces. Indeed,

$$J = \begin{pmatrix} 0 & 0 & 0 & 1 \\ 0 & 0 & I & 0 \\ 0 & I & 0 & 0 \\ 1 & 0 & 0 & 0 \end{pmatrix}$$

is an Hermitian unitary operator on the Hilbert space $\mathcal{K} = \mathbb{C} \times \mathcal{H} \times \mathcal{H} \times \mathbb{C}$, and so we can define a Pontrjagin space $\mathcal{K}_J$ with indefinite inner product defined by J:

$$\begin{aligned} < x, y >_J &= (Jx|y) \\ &= \alpha_4\bar{\beta}_1 + (x_3|y_2) + (x_2|y_3) + \alpha_1\bar{\beta}_4. \end{aligned}$$

Then, an element A of $T^*(\mathcal{H})$ is regarded as an operator on the Pontrjagin space $\mathcal{K}_J$:

$$\begin{pmatrix} 0 & \lambda^*(A) & 0 & 0 \\ 0 & \pi(A) & 0 & 0 \\ 0 & 0 & \pi(A) & \lambda(A) \\ 0 & 0 & 0 & 0 \end{pmatrix}.$$

Notes For Banach $*$-algebras, C^*-algebras and the algebraic notions ($*$-homomorphisms, $*$-isomorphisms, etc.), refer to Appendix C. As for Krein spaces, refer to [23].

2.2 Structure of CQ^*-algebras

In this section we show that a CQ^*-algebra is divided into two types by its expectation, and that the study of CQ^*-algebras results in that of CT^*-algebras.

Lemma 2.2.1 *Let $\mathfrak{A}$ be a Q^*-algebra (or, a T^*-algebra) on a Hilbert space $\mathcal{H}$. If the map: $\pi(A) \to \lambda(A)$ is uniformly continuous, then there exists an element g of $\mathcal{H}$ such that $\lambda(A) = \pi(A)g$ for all $A \in \mathfrak{A}$.*

Proof Since $\pi(A) \to \lambda(A)$ is uniformly continuous, there exists a constant $\gamma > 0$ such that

$$\|\lambda(A)\| \leqq \gamma \|\pi(A)\| \tag{2.2.1}$$

for all $A \in \mathfrak{A}$. For any $\varepsilon > 0$ and $A_1, A_2, \ldots, A_n \in \mathfrak{A}$, we put

$$\begin{aligned}&\Delta(\varepsilon; A_1, \ldots, A_n)\\&= \{x \in \mathcal{H};\ \|x\| \leqq \gamma,\ \|\pi(A_k)x - \lambda(A_k)\| \leqq \varepsilon \ \text{ for all } k = 1, \ldots, n\}.\end{aligned}$$

Since the uniform closure of $\pi(\mathfrak{A})$ in $B(\mathcal{H})$ is a C^*-algebra on $\mathcal{H}$, it has an approximate identity, so that there exists an element A of $\mathfrak{A}$ such that

$$\begin{aligned}&\|\pi(A)\| \leqq 1,\\&\|\pi(A_k)\pi(A) - \pi(A_k)\| < \frac{\varepsilon}{\gamma}, \quad k = 1, 2, \ldots, n.\end{aligned} \tag{2.2.2}$$

Then it follows from (2.2.1) and (2.2.2) that

$$\begin{aligned}\|\lambda(A)\| &\leqq \gamma,\\ \|\pi(A_k)\lambda(A) - \lambda(A_k)\| &= \|\lambda(A_k A - A_k)\|\\ &\leqq \gamma \|\pi(A_k)\pi(A) - \pi(A_k)\|\\ &< \varepsilon\end{aligned}$$

for $k = 1, 2, \ldots, n$. Hence, $\lambda(A) \in \Delta(\varepsilon; A_1, \ldots, A_n)$, so that $\Delta(\varepsilon; A_1, \ldots, A_n)$ is a nonempty weakly closed subset in the weakly compact set $\mathcal{H}_\gamma := \{x \in \mathcal{H};\ \|x\| \leqq \gamma\}$; hence it is weakly compact. Therefore, we can take an element g of $\bigcap\{\Delta(\varepsilon; A);\ \varepsilon > 0,\ A \in \mathfrak{A}\}$, which implies that $\lambda(A) = \pi(A)g$ for all $A \in \mathfrak{A}$. This completes the proof.

Lemma 2.2.2 *Let $\mathfrak{A}$ be a Q^*-algebra on $\mathcal{H}$ and μ its expectation. Then the following (i)–(v) are equivalent:*

(i) $\pi(A) \to \mu(A)$ *is a continuous linear functional on the normed space* $\pi(\mathfrak{A})$*, that is,*

$$\|\mu\|_\pi := \sup\{|\mu(A)|; \quad A \in u_\pi(\mathfrak{A})\} < \infty,$$

where $u_\pi(\mathfrak{A}) := \{A \in \mathfrak{A}; \quad \|\pi(A)\| \leqq 1\}$.

(ii) $\tau(A) \to \mu(A)$ *is a continuous linear functional on the normed space* $\tau(\mathfrak{A})$*, that is,*

$$\|\mu\|_\tau := \sup\{|\mu(A)|; \quad A \in u(\tau(\mathfrak{A}))\} < \infty,$$

where $u(\tau(\mathfrak{A})) := \{A \in \mathfrak{A}; \quad \|\tau(A)\| \leqq 1\}$.

(iii) There exists an element g *of* $\mathcal{H}$ *such that*

$$A = (\pi(A), \pi(A)g, (\pi(A)^* g)^*, (\pi(A)g|g))$$

for all $A \in \mathfrak{A}$.

(iv) $\mathfrak{A}$ *is homeomorphic to* $\pi(\mathfrak{A})$ *as normed space.*

Furthermore, if $\mathfrak{A}$ *is a* CQ^**-algebra on* $\mathcal{H}$*, then above equivalent statements (i)–(iv) are equivalent to the following (v):*

(v) $\pi(\mathfrak{A})$ *is a* C^**-algebra on* $\mathcal{H}$ *and* $\pi(A) \to \mu(A)$ *is a mapping.*

Proof

(iv)⇒(i)⇒(ii) This is trivial.

(ii)⇒(iii) By (ii) there exists a constant $\gamma > 0$ such that

$$|\mu(A)| \leqq \gamma \max(\|\pi(A)\|, \|\lambda(A)\|, \|\lambda(A^\sharp)\|) \tag{2.2.3}$$

for all $A \in \mathfrak{A}$. Hence we have

$$\begin{aligned}\|\lambda(A)\|^2 &= \mu(A^\sharp A)\\ &\leqq \gamma \max(\|\pi(A^\sharp A)\|, \|\lambda(A^\sharp A)\|)\\ &\leqq \gamma \max(\|\pi(A)\|^2, \|\pi(A)\|\|\lambda(A)\|),\end{aligned}$$

which implies that

$$\|\lambda(A)\| \leqq \max(\gamma, 1)\|\pi(A)\|$$

for all $A \in \mathfrak{A}$. By Lemma 2.2.1 there exists an element g of $\mathcal{H}$ such that $\lambda(A) = \pi(A)g$ for all $A \in \mathfrak{A}$, which yields by (2.2.3) that

$$|\mu(A)| \leqq (\gamma + \|g\|)\|\pi(A)\| \tag{2.2.4}$$

for all $A \in \mathfrak{A}$. Take an arbitrary $A \in \mathfrak{A}$. As shown in (2.2.2), for any $\varepsilon > 0$ we can find an element B of $\mathfrak{A}$ such that

$$\|\pi(B)\pi(A) - \pi(A)\| < \varepsilon$$

Hence it follows from (2.2.4) that

$$\begin{aligned}
|(\pi(A)g|g) - \mu(A)| &\leqq |(\pi(A)g|g) - (\pi(B)\pi(A)g|g)| \\
&\quad + |(\pi(B)\pi(A)g|g) - \mu(A)| \\
&= |((\pi(B)\pi(A) - \pi(A))g|g)| \\
&\quad + |\mu(BA) - \mu(A)| \\
&\leqq \|\pi(B)\pi(A) - \pi(A)\|\|g\|^2 \\
&\quad + \max(\gamma, \|g\|)\|\pi(B)\pi(A) - \pi(A)\| \\
&< (\|g\|^2 + \max(\gamma, \|g\|))\varepsilon,
\end{aligned}$$

so that $\mu(A) = (\pi(A)g|g)$, which proves (iii).

(iii)⇒(iv) This follows from

$$\|\pi(A)\| \leqq \|A\| \leqq \max(1, \|g\|^2)\|\pi(A)\|$$

for all $A \in \mathfrak{A}$. Thus, (i)–(iv) are equivalent.

Suppose that $\mathfrak{A}$ is a CQ^*-algebra on $\mathcal{H}$.

(iv)⇒(v) This is trivial.

(v)⇒(i) Putting

$$\tilde{\mu}(\pi(A)) = \mu(A), \quad A \in \mathfrak{A},$$

$\tilde{\mu}$ is a positive linear functional on the C^*-algebra $\pi(\mathfrak{A})$. Hence it follows from [37, Proposition 9.12] that $\tilde{\mu}$ is continuous, which derives (i). This completes the proof.

Lemma 2.2.3 *Let $\mathfrak{A}$ be a CQ^*-algebra on $\mathcal{H}$ and μ its expectation. If the linear functional on the normed $*$-algebra $\tau(\mathfrak{A})$ defined by $\tau(A) \rightarrow \mu(A)$ is not continuous, then $\mathbb{C}V$ is a closed $*$-ideal of $\mathfrak{A}$, and $\tau(\mathfrak{A})$ is a CT^*-algebra on $\mathcal{H}$ which is isomorphic to $\mathfrak{A}/\mathbb{C}V$.*

Proof Since $\tau(A) \to \mu(A)$ is not continuous, we can choose a sequence $\{B_n\}$ in $\mathfrak{A}$ such that $\lim_{n\to\infty} \|\tau(B_n)\| = 0$ and $\mu(B_n) \geqq 1$, $n = 1, 2, \ldots$. We here put

$$A_n = \frac{B_n}{\mu(B_n)}, \quad n = 1, 2, \ldots.$$

Then $\{A_n\} \subset \mathfrak{A}$, $\lim_{n\to\infty} \|\tau(A_n)\| = 0$ and $\mu(A_n) = 1$, $n = 1, 2, \ldots$. Therefore,

$$\lim_{n\to\infty} A_n = \lim_{n\to\infty} (\tau(A_n), \mu(A_n)) = V.$$

Since $\mathfrak{A}$ is a CQ^*-algebra, it follows that $V \in \mathfrak{A}$ and $\mathbb{C}V$ is a closed $*$-ideal of $\mathfrak{A}$. Hence, $\tau(\mathfrak{A}) := \{(\pi(A), \lambda(A), \lambda^*(A));\ A \in \mathfrak{A}\}$ is isomorphic to $\mathfrak{A}/_{\mathbb{C}V}$ as a Banach $*$-algebra. Thus, $\tau(\mathfrak{A})$ is a CT^*-algebra on $\mathcal{H}$. This completes the proof.

By Lemmas 2.2.2 and 2.2.3 we have the following

Theorem 2.2.4 *Let $\mathfrak{A}$ be a CQ^*-algebra on $\mathcal{H}$ and μ its expectation. Then*

(1) $\|\mu\|_\pi$ is finite if and only if there exists an element g of $\mathcal{H}$ such that

$$A = (\pi(A), \pi(A)g, (\pi(A)^*g)^*, (\pi(A)g|g))$$

for all $A \in \mathfrak{A}$,

(2) $\|\mu\|_\pi$ is infinite if and only if $\mathbb{C}V$ is a closed $*$-ideal of $\mathfrak{A}$ and the CT^*-algebra $\tau(\mathfrak{A})$ is isomorphic to the CT^*-algebra $\mathfrak{A}/_{\mathbb{C}V}$.

In case (1) in Theorem 2.2.4 a CQ^*-algebra $\mathfrak{A}$ is decided by $\pi(\mathfrak{A})$, and in case (2) it is decided by the CT^*-algebra $\tau(\mathfrak{A})$.

2.3 Functional Calculus for Self-adjoint Trio Observables

We begin with a spectrum of trio observable and a functional calculus of a self-adjoint trio observable. For $A \in T^*(\mathcal{H})$ we denote by $CT^*(A)$ the CT^*-algebra generated by A. Let $T^*_{\mathbb{1}}(\mathcal{H})$ denote the adjunction of identity $\mathbb{1}$ to $T^*(\mathcal{H})$, and it is also called the *unitization* of $T^*(\mathcal{H})$. We define the spectrum of $A \in T^*(\mathcal{H})$ as follows:

$$S_p(A) = \{\lambda \in \mathbb{C};\ (A - \lambda\mathbb{1})^{-1} \text{ does not exist in } T^*_{\mathbb{1}}(\mathcal{H})\}.$$

Proposition 2.3.1 *Let $A \in T^*(\mathcal{H})$. Then*

$$S_p(A) = S_p(\pi(A)) \cup \{0\},$$

where $S_p(\pi(A))$ is the spectrum of $\pi(A)$ defined by

$$S_p(\pi(A)) = \{\lambda \in \mathbb{C}; \quad (\pi(A) - \lambda I)^{-1} \textit{ does not exist in } B(\mathcal{H})\}.$$

Proof Suppose $0 \notin S_p(A)$. Then there exist $\alpha \in \mathbb{C}$ and $B \in T^*(\mathcal{H})$ such that $(B + \alpha \mathbb{1})A = \mathbb{1}$, which implies that $\mathbb{1} = BA + \alpha A \in T^*(\mathcal{H})$. This contradicts that $T^*(\mathcal{H})$ does not have identity. Hence we have

$$0 \in S_p(A). \tag{2.3.1}$$

Next, take an arbitrary non-zero $\lambda \notin S_p(A)$. Then there exist $\alpha \in \mathbb{C}$ and $B \in T^*(\mathcal{H})$ such that

$$(B - \alpha \mathbb{1})(A - \lambda \mathbb{1}) = (A - \lambda \mathbb{1})(B - \alpha \mathbb{1}) = \mathbb{1},$$

so that

$$\alpha\lambda = 1 \text{ and } BA - \lambda B - \frac{1}{\lambda}A = AB - \frac{1}{\lambda}A - \lambda B = O,$$

which implies that

$$(\pi(B) - \frac{1}{\lambda}I)(\pi(A) - \lambda I) = (\pi(A) - \lambda I)(\pi(B) - \frac{1}{\lambda}I) = I.$$

This means that $(\pi(A) - \lambda I)^{-1} = (\pi(B) - \frac{1}{\lambda}I)$. Thus, $\lambda \notin S_p(\pi(A))$, so we get

$$S_p(\pi(A)) \subset S_p(A). \tag{2.3.2}$$

Conversely, take an arbitrary non-zero $\lambda \notin S_p(\pi(A))$. Then there exists an element B_0 of $B(\mathcal{H})$ such that

$$(B_0 - \frac{1}{\lambda}I)(\pi(A) - \lambda I) = (\pi(A) - \lambda I)(B_0 - \frac{1}{\lambda}I) = I.$$

We here write

$$x = \frac{1}{\lambda}(B_0\lambda(A) - \frac{1}{\lambda}\lambda(A)),$$
$$y = \frac{1}{\lambda}((\pi(A) - \lambda I)^{-1})^*\lambda(A^\sharp)),$$

and

$$B = (B_0, x, y^*).$$

Then it follows that $B \in T^*(\mathcal{H})$ and

$$(B - \frac{1}{\lambda}\mathbb{1})(A - \lambda\mathbb{1}) = (A - \lambda\mathbb{1})(B - \frac{1}{\lambda}\mathbb{1}) = \mathbb{1},$$

which yields that $\lambda \notin S_p(A)$. Therefore

$$S_p(A) \subset S_p(\pi(A)) \cup \{0\}. \tag{2.3.3}$$

By (2.3.1)–(2.3.3) we have

$$S_p(A) = S_p(\pi(A)) \cup \{0\}.$$

This completes the proof.

We first consider a functional calculus of self-adjoint trio observables. Let $A = A^\sharp \in T^*(\mathcal{H})$. Then $\pi(A)^* = \pi(A)$ and $\lambda^*(A) = \lambda(A)^*$. By Proposition 2.3.1 and Appendix A.1 $S_p(A)$ is a compact subset of $\mathbb{R}$ and is contained in $[-\|\pi(A)\|, \|\pi(A)\|]$. For any polynomial on $\mathbb{R}$:

$$p(t) = \alpha_0 + \alpha_1 t + \cdots + \alpha_n t^n,$$

we define an element $p(A)$ of $T^*_{\mathbb{1}}(\mathcal{H})$ as follows:

$$p(A) = \alpha_0 \mathbb{1} + \alpha_1 A + \cdots + \alpha_n A^n.$$

Then we have

$$\begin{aligned} p(A) &= \alpha_0 \mathbb{1} + (\alpha_1 \pi(A) + \cdots + \alpha_n \pi(A)^n, (\alpha_1 I + \alpha_2 \pi(A) \\ &\quad + \cdots + \alpha_n \pi(A)^{n-1})\lambda(A), ((\bar{\alpha}_1 I + \bar{\alpha}_2 \pi(A) + \cdots + \bar{\alpha}_n \pi(A)^{n-1})\lambda(A))^*) \\ &= p(0)\mathbb{1} + (p_\pi(\pi(A)), p_\lambda(\pi(A))\lambda(A), (\bar{p}_\lambda(\pi(A))\lambda(A))^*), \end{aligned}$$

where

$$p_\pi(t) := p(t) - p(0),$$

$$p_\lambda(t) := \alpha_1 + \alpha_2 t + \cdots + \alpha_n t^{n-1} = \begin{cases} \frac{p(t)-p(0)}{t}, & t \neq 0, \\ p'(0), & t = 0, \end{cases} \tag{2.3.4}$$

$$\overline{p_\lambda}(t) := \overline{p_\lambda(t)}.$$

It is clear that $p(A) \in T^*(\mathcal{H})$ if and only if $p(0) = 0$. Based on this fact, we consider a functional calculus for a Banach $*$-algebra $C_\delta(S_p(A))$ of continuous functions on $S_p(A)$ defined below:

Let f be a continuous function on $S_p(A)$ vanishing at 0 and having the formal derivative $f'(0)$ at 0. Then we have

$$\|f\|_u := \sup_{t \in S_p(A)} |f(t)| = \|f(\pi(A))\|. \tag{2.3.5}$$

As the same as (2.3.4), we define a Borel function f_λ on $S_p(A)$ by

$$f_\lambda(t) = \begin{cases} \frac{f(t)}{t}, & t \neq 0 \\ f'(0), & t = 0. \end{cases}$$

Then if $f_\lambda(\pi(A))\lambda(A)$ is well-defined, then

$$\|f_\lambda\|_\mu := \|f_\lambda(\pi(A))\lambda(A)\| = \left[\int |f_\lambda(t)|^2 d\mu_A(t)\right]^{\frac{1}{2}} < \infty, \tag{2.3.6}$$

where $\pi(A) = \int t dE(t)$ is the spectral resolution of $\pi(A)$, and μ_A is the bounded Borel measure on $\mathbb{R}$ defined by $\mu_A(t) = (E(t)\lambda(A)|\lambda(A))$ (see Appendix A.4). However, we don't know whether $f_\lambda(\pi(A))\lambda(A)$ is well-defined in general. We now denote by $C_\delta(S_p(A))$ the set of all continuous functions f on $S_p(A)$ with formal derivative $f'(0)$ at 0 such that $f(0) = 0$ and $\|f_\lambda\|_\mu < \infty$. For any $f \in C_\delta(S_p(A))$ we can define an element $f(A)$ of $T^*(\mathcal{H})$ as follows:

$$f(A) = \left(f(\pi(A)), f_\lambda(\pi(A))\lambda(A), (\bar{f}_\lambda(\pi(A))\lambda(A))^*\right). \tag{2.3.7}$$

It is easily shown that $C_\delta(S_p(A))$ is a Banach space under the usual function operations $f + g, \alpha f$ and the norm:

$$\|f\| := \max(\|f\|_u, \|f_\lambda\|_\mu). \tag{2.3.8}$$

Furthermore, since

$$(fg)_\lambda = fg_\lambda \quad \text{and} \quad (f^*)_\lambda = (\bar{f})_\lambda, \tag{2.3.9}$$

where $\bar{f}(t) = \overline{f(t)}, t \in S_p(A)$, we have

$$\begin{aligned} \|fg\|_u &\leqq \|f\|_u\|g\|_u \leqq \|f\|\|g\|, \\ \|(fg)_\lambda\|_\mu &= \|f(\pi(A))g_\lambda(\pi(A))\lambda(A)\| \\ &\leqq \|f\|_u\|g_\lambda\|_\mu \end{aligned}$$

$$\leqq \|f\|\|g\|,$$
$$\|f^*\|_u = \|f\|_u \quad \text{and} \quad \|(f_\lambda)^*\|_\mu = \|f_\lambda\|_\mu,$$

which shows that $C_\delta(S_p(A))$ is a Banach $*$-algebra. Furthermore, it follows from (2.3.5)–(2.3.8) that

$$\begin{aligned}\|f(A)\| &= \max(\|f(\pi(A))\|, \|f_\lambda(\pi(A))\lambda(A)\|) \\ &= \max\left(\|f\|_u, \|f_\lambda\|_\mu\right) = \|f\|\end{aligned} \tag{2.3.10}$$

for all $f \in C_\delta(S_p(A))$.

The following theorem for the functional calculus for $C_\delta(S_p(A))$ is one of the main results in this section.

Theorem 2.3.2 *Let $A = A^\sharp \in T^*(\mathcal{H})$. For any $f \in C_\delta(S_p(A))$ we put*

$$f(A) = \left(f(\pi(A)), f_\lambda(\pi(A))\lambda(A), (\bar{f}_\lambda(\pi(A))\lambda(A))^*\right).$$

Then the map: $f \to f(A)$ is a $*$-isomorphism of the Banach $*$-algebra $C_\delta(S_p(A))$ onto the Banach $*$-algebra $CT^*(A)$, that is, for any $f,\ g \in C_\delta(S_p(A))$ and $\alpha \in \mathbb{C}$ we have

$$\begin{aligned}(f+g)(A) &= f(A) + g(A), \\ (\alpha f)(A) &= \alpha f(A), \\ (fg)(A) &= f(A)g(A), \\ f^*(A) &= f(A)^\sharp, \\ \|f\| &= \|f(A)\|.\end{aligned}$$

To prove this theorem, we prepare some statements. Let $\alpha \notin S_p(A)$ and put

$$q_\alpha(t) = t(t-\alpha)^{-1}, \quad t \in S_p(A).$$

Then it is clear that $q_\alpha \in C_\delta(S_p(A))$.

Lemma 2.3.3 *For any $\alpha \notin S_p(A)$, $q_\alpha(A) \in CT^*(A)$.*

Proof It is clear that if $|\alpha| > \|\pi(A)\|$, then q_α is analytic on $S_p(A)$ and $q_\alpha(t) = -\sum_{n=1}^{\infty}(\frac{t}{\alpha})^n$. Hence, $q_\alpha(A) \in CT^*(A)$. Take an arbitrary $\alpha \notin S_p(A)$ such that $q_\alpha(A) \in CT^*(A)$. Then we have

$$\alpha q_\alpha(t) - \beta q_\beta(t) = (\alpha - \beta) q_\alpha(t) q_\beta(t), \quad t \in S_p(A) \tag{2.3.11}$$

for each $\beta \notin S_p(A)$, which implies $q_\beta(A) \in CT^*(A)$ for some neighborhood of α. Furthermore, since $\mathbb{C} - S_p(A)$ is connected, it follows from the analytic continuation that $q_\alpha(A) \in CT^*(A)$ for all $\alpha \notin S_p(A)$.

Let $\mathfrak{M}$ be the closed subspace of $C_\delta(S_p(A))$ generated by $\{q_\alpha; \alpha \notin S_p(A)\}$. By (2.3.11) and Lemma 2.3.3, we can prove

$$q_\alpha q_\beta = \frac{1}{\alpha - \beta}(\alpha q_\alpha - \beta q_\beta) \in \mathfrak{M}$$

for any α, $\beta \notin S_p(A)$ with $\alpha \neq \beta$. Hence $\mathfrak{M}$ is a closed $*$-subalgebra of the Banach $*$-algebra $C_\delta(S_p(A))$. We will verify that $\mathfrak{M} = C_\delta(S_p(A))$ under some preparations.

Lemma 2.3.4 *Suppose that f is an analytic function on $S_p(A)$ such that $f(0) = 0$. Then $f \in \mathfrak{M}$.*

Proof Since $g(t) := \frac{f(t)}{t}$ is an analytic function on $S_p(A)$ by $f(0) = 0$, we have

$$\begin{aligned} f(t) = tg(t) &= \frac{1}{2\pi i} \int_{|\alpha - t| = 1} \frac{tg(\alpha)}{\alpha - t} d\alpha \\ &= -\frac{1}{2\pi i} \int_{|\alpha - t| = 1} g(\alpha) q_\alpha(t) d\alpha \\ &\in \mathfrak{M}. \end{aligned}$$

Lemma 2.3.5 *For any $f \in C_\delta(S_p(A))$ and $\alpha \notin S_p(A)$ we have $f q_\alpha \in \mathfrak{M}$.*

Proof For any polynomial $p(t) = \alpha_0 + \alpha_1 t + \cdots + \alpha_n t^n$, we can prove $pq_\alpha \in \mathfrak{M}$. Indeed, it follows from Lemma 2.3.4 that $r(t) := \alpha_1 t + \cdots + \alpha_n t^n \in \mathfrak{M}$, which implies since $\mathfrak{M}$ is an algebra that $pq_\alpha = \alpha_0 q_\alpha + r q_\alpha \in \mathfrak{M}$. Take an arbitrary $f \in C_\delta(S_p(A))$. By the Weierstrass approximation theorem, there exists a sequence $\{p_n\}$ of polynomials which converges uniformly to f on $S_p(A)$. Then, since $\{p_n q_\alpha\}$ converges uniformly to $f q_\alpha$ on $S_p(A)$ and $p_n q_\alpha$, $f q_\alpha \in C_\delta(S_p(A))$, $n = 1, 2, \ldots$, we have

$$\lim_{n \to \infty} \|p_n q_\alpha - f q_\alpha\|_u = 0.$$

Furthermore, the equalities:

$$(p_n q_\alpha)_\lambda = (p_n)_\lambda q_\alpha + p_n(0)(q_\alpha)_\lambda$$

and

$$(f q_\alpha)_\lambda = f_\lambda q_\alpha,$$

implies that

$$\begin{aligned}\|(p_n q_\alpha)_\lambda - (f q_\alpha)_\lambda\|_\mu &= \|\left((p_n)_\lambda(\pi(A)) - f_\lambda(\pi(A))\right) q_\alpha(\pi(A))\lambda(A) \\ &\quad + p_n(0)(q_\alpha)_\lambda(\pi(A))\lambda(A)\| \\ &= \|\left(p_n(\pi(A)) - f(\pi(A)) - p_n(0)\right)(\pi(A)-\alpha I)^{-1}\lambda(A) \\ &\quad + p_n(0)(q_\alpha)_\lambda(\pi(A))\lambda(A)\| \\ &\leqq \|\left(p_n(\pi(A)) - f(\pi(A))\right)(\pi(A)-\alpha I)^{-1}\lambda(A)\| \\ &\quad + |p_n(0)| \left\{ \|(\pi(A)-\alpha I)^{-1}\lambda(A)\| + \|(q_\alpha)_\lambda(\pi(A))\lambda(A)\| \right\} \\ &\to 0 \quad \text{as} \quad n \to \infty.\end{aligned}$$

Therefore

$$\lim_{n\to\infty} \|p_n q_\alpha - f q_\alpha\| = 0.$$

Furthermore, since $\{p_n q_\alpha\} \subset \mathfrak{M}$, we obtain $f q_\alpha \in \mathfrak{M}$.

The Proof of Theorem 2.3.2 Take an arbitrary $f \in C_\delta(S_p(A))$. Since $q_\alpha^{-1} f$ is continuous on $S_p(A) \setminus \{0\}$, we can choose a sequence $\{f_n\}$ of continuous functions on $S_p(A)$ satisfying the following

(1) $f_n(0) = 0$,
(2) $f_n(t) = (q_\alpha^{-1} f)(t), \quad |t| \geqq \frac{1}{n}$,
(3) $\sup_{t\in S_p(A)} |(f_n q_\alpha)_\lambda(t) - f_\lambda(t)| \leqq \gamma$ for some constant $\gamma > 0$.

Then we have

$$\begin{aligned}\sup_{t\in S_p(A)} |(f_n q_\alpha)(t) - f(t)| &= \sup_{-\frac{1}{n}\leqq t\leqq \frac{1}{n}} |((f_n q_\alpha)_\lambda(t) - f_\lambda(t))t| \\ &\leqq \gamma \frac{1}{n} \to 0 \quad \text{as} \quad n \to \infty.\end{aligned}$$

Furthermore, since $(f_n q_\alpha)_\lambda$ converges pointwise to f_λ and $\|f_\lambda\|_\mu < \infty$, we have

$$\begin{aligned}\|(f_n q_\alpha)_\lambda - f_\lambda\|_\mu &= \left[\int |(f_n q_\alpha)_\lambda(t) - f_\lambda(t)|^2 d\mu_A(t)\right]^{\frac{1}{2}} \\ &\to 0 \quad \text{as} \quad n \to \infty.\end{aligned}$$

Hence it follows that $\{f_n q_\alpha\} \subset \mathfrak{M}$ and $\lim_{n\to\infty} \|f_n q_\alpha - f\| = 0$, which implies $f \in \mathfrak{M}$. Thus $\mathfrak{M} = C_\delta(S_p(A))$. By Lemma 2.3.3 and (2.3.10), we have

$$\{f(A);\ f \in C_\delta(S_p(A))\} = CT^*(A),$$

so that the map:

$$f \in C_\delta(S_p(A)) \to f(A) \in CT^*(A)$$

is surjective. Furthermore, using the functional calculus theorem for continuous functions of $\pi(A)$ [Theorem A.1 in Appendix A], we can prove that the map: $f \to f(A)$ is a $*$-isomorphism of the Banach $*$-algebra $C_\delta(S_p(A))$ onto the Banach $*$-algebra $CT^*(A)$. Indeed, it is clear that the map $f \to f(A)$ is linear and $*$-preserving. Furthermore, for any $f,\ g \in C_\delta(S_p(A))$ it follows from (2.3.9) and (2.3.10) that

$$\begin{aligned}(fg)(A) &= \big((fg)(\pi(A)), (fg)_\lambda(\pi(A))\lambda(A), ((fg)_\lambda^*(\pi(A))\lambda(A))^*\big)\\ &= \big(f(\pi(A))g(\pi(A)), f(\pi(A))g_\lambda(\pi(A))\lambda(A), (\bar{g}(\pi(A))\bar{f}_\lambda(\pi(A))\lambda(A))^*\big)\\ &= f(A)g(A),\end{aligned}$$

and

$$\begin{aligned}\|f\| &= \max(\|f\|_u, \|f_\lambda\|_\mu)\\ &= \max(\|f(\pi(A))\|, \|f_\lambda(\pi(A))\lambda(A)\|)\\ &= \|f(A)\|.\end{aligned}$$

This completes the proof.

In this paper we denote by $[\mathcal{K}]$ the closed subspace generated by a subset $\mathcal{K}$ of the Hilbert space $\mathcal{H}$ under the norm topology on $\mathcal{H}$, and denote by Proj $[\mathcal{K}]$ the projection on the closed subspace $[\mathcal{K}]$ of $\mathcal{H}$. Furthermore, we denote by $\ker X_0$ and $R(X_0)$ the kernel and the range of $X_0 \in B(\mathcal{H})$, respectively, namely $\ker X_0 := \{x \in \mathcal{H};\ X_0x = 0\}$ and $R(X_0) := \{X_0x;\ x \in \mathcal{H}\}$. By Theorem 2.3.2 we get a decomposition theorem of a self-adjoint trio observable as follows:

Corollary 2.3.6 *Let $A = A^\sharp \in T^*(\mathcal{H})$ and P_A the projection onto $[R(\pi(A))]$. We define elements A_s and A_n of $T^*(\mathcal{H})$ by*

$$\begin{aligned}A_s &= (\pi(A), P_A\lambda(A), (P_A\lambda(A))^*),\\ A_n &= (0, (I - P_A)\lambda(A), ((I - P_A)\lambda(A))^*).\end{aligned}$$

Then

$$A_s,\ A_n \in CT^*(A) \quad and \quad A = A_s + A_n.$$

Proof Define a function f on $S_p(A)$ and a formal derivative $f'(0)$ at 0 as follows:

$$f(t) = 0 \quad \text{on} \quad S_p(A) \quad \text{and} \quad f'(0) = 1.$$

Then,

$$f_\lambda(t) = \begin{cases} 0, & t \neq 0 \\ 1, & t = 0. \end{cases}$$

Let $\pi(A) = \int t dE(t)$ be the spectral resolution of $\pi(A)$. Then since $f(\pi(A)) = 0$ and

$$\begin{aligned} f_\lambda(\pi(A)) &= \int_{\{0\}} t dE(t) \\ &= \text{Proj } \ker \pi(A) \\ &= \text{Proj } R(\pi(A))^\perp \\ &= I - P_A, \end{aligned}$$

where $\mathcal{K}^\perp$ denotes the orthogonal complement of a subset $\mathcal{K}$ of $\mathcal{H}$, it follows that $f \in C_\delta(S_p(A))$, which implies by Theorem 2.3.2 that

$$\begin{aligned} f(A) &= \left(f(\pi(A)), f_\lambda(\pi(A))\lambda(A), (\bar{f_\lambda}(\pi(A))\lambda(A))^*\right) \\ &= (0, (I - P_A)\lambda(A), ((I - P_A)\lambda(A))^*) \\ &= A_n \in CT^*(A). \end{aligned}$$

Hence,

$$\begin{aligned} A_s &= (\pi(A), P_A\lambda(A), (P_A\lambda(A))^*) \\ &= A - A_n \in CT^*(A). \end{aligned}$$

This completes the proof.

The element A_s in Corollary 2.3.6 is called the *semisimple part* of A and A_n is called the *nilpotent part* of A. If $A = A_s$, then A is called *semisimple*, and if $A = A_n$, then A is called *nilpotent*.

We next discuss a square root and a polar decomposition of $A \in T^*(\mathcal{H})$ using Theorem 2.3.2. The following theorem shows that we can define the square root of $A^\sharp A$. Let

$$\pi(A) = U_A(\pi(A)^*\pi(A))^{\frac{1}{2}}$$

be the polar decomposition of $\pi(A)$. Here U_A is a unique partial isometry on $\mathcal{H}$ such that $U_A^*U_A = \text{Proj}\left[(\pi(A)^*\pi(A))^{\frac{1}{2}}\mathcal{H}\right]$ and $U_AU_A^* = \text{Proj}\left[(\pi(A)\pi(A)^*)^{\frac{1}{2}}\mathcal{H}\right]$ (see Theorem A.2 in Appendix A). Then

$$\pi(A) = (\pi(A)\pi(A)^*)^{\frac{1}{2}}U_A.$$

Using Theorem 2.3.2, we can prove a fundamental result in this note.

Theorem 2.3.7 *Let $A \in T^*(\mathcal{H})$. We define elements $(AA^\sharp)^{\frac{1}{2}}$ and $(A^\sharp A)^{\frac{1}{2}}$ of $T^*(\mathcal{H})$ by*

$$(AA^\sharp)^{\frac{1}{2}} = \left((\pi(A)\pi(A)^*)^{\frac{1}{2}}, U_A\lambda(A^\sharp), (U_A\lambda(A^\sharp))^*\right)$$

and

$$(A^\sharp A)^{\frac{1}{2}} = \left((\pi(A)^*\pi(A))^{\frac{1}{2}}, U_A^*\lambda(A), (U_A^*\lambda(A))^*\right).$$

Then they belong to $CT^*(A)$ and they satisfy that $((AA^\sharp)^{\frac{1}{2}})^2 = AA^\sharp$ and $((A^\sharp A)^{\frac{1}{2}})^2 = A^\sharp A$.

Proof Putting

$$f(t) = t^{\frac{1}{2}}, \quad t \geqq 0,$$

we see that

$$f_\lambda(t) = \begin{cases} t^{-\frac{1}{2}}, & t > 0 \\ c, & t = 0, \end{cases}$$

where c is any constant. Therefore

$$f(\pi(AA^\sharp)) = f(\pi(A)\pi(A)^*) = (\pi(A)\pi(A)^*)^{\frac{1}{2}}$$

and

$$\begin{aligned} f_\lambda(\pi(AA^\sharp))\lambda(AA^\sharp) &= f_\lambda(\pi(A)\pi(A)^*)\pi(A)\lambda(A^\sharp) \\ &= f_\lambda(\pi(A)\pi(A)^*)(\pi(A)\pi(A)^*)^{\frac{1}{2}}U_A\lambda(A^\sharp) \\ &= U_A\lambda(A^\sharp), \end{aligned}$$

so that $f \in C_\delta(S_p(AA^\sharp))$, which implies by Theorem 2.3.2 that $f(AA^\sharp) \in CT^*(AA^\sharp)$, $f(AA^\sharp) = (AA^\sharp)^{\frac{1}{2}}$ and $f(AA^\sharp)^2 = AA^\sharp$. Similarly, we can prove that $(A^\sharp A)^{\frac{1}{2}} \in CT^*(A)$ and $((A^\sharp A)^{\frac{1}{2}})^2 = A^\sharp A$. This completes the proof.

The elements $(AA^\sharp)^{\frac{1}{2}}$ and $(A^\sharp A)^{\frac{1}{2}}$ of $CT^*(A)$ in Theorem 2.3.7 are called the *square roots* of $AA^\sharp$ and $A^\sharp A$, respectively. It is natural to consider the following question: *Is the square root $(A^\sharp A)^{\frac{1}{2}}$ of $A^\sharp A$ unique?* For this question we obtain the following

Proposition 2.3.8 *Let $A \in T^*(\mathcal{H})$. Then, the square root $(A^\sharp A)^{\frac{1}{2}}$ of $A^\sharp A$ is unique (in the sense that if $A^\sharp A = B^\sharp B$ for $B \in T^*(\mathcal{H})$, then $(A^\sharp A)^{\frac{1}{2}} = (B^\sharp B)^{\frac{1}{2}}$), if and only if* $\ker \pi(A) = \{0\}$.

Proof Let $A,\ B \in T^*(\mathcal{H})$. Then we have

$$\begin{aligned} A^\sharp A = B^\sharp B \quad &\text{if and only if} \quad \pi(A)^*\pi(A) = \pi(B)^*\pi(B) \\ &\qquad \text{and} \quad \pi(A)^*\lambda(A) = \pi(B)^*\lambda(B) \\ &\text{if and only if} \quad |\pi(A)| := (\pi(A)^*\pi(A))^{\frac{1}{2}} = (\pi(B)^*\pi(B))^{\frac{1}{2}} = |\pi(B)| \\ &\qquad \text{and } |\pi(A)|U_A^*\lambda(A) = |\pi(B)|U_B^*\lambda(B) = |\pi(A)|U_B^*\lambda(B) \\ &\text{if and only if} \quad |\pi(A)| = |\pi(B)| \\ &\qquad \text{and } U_A^*\lambda(A) - U_B^*\lambda(B) \in \ker|\pi(A)| = \ker \pi(A). \end{aligned} \tag{2.3.12}$$

Suppose that $\ker \pi(A) = \{0\}$ and $A^\sharp A = B^\sharp B$ for $B \in T^*(\mathcal{H})$. Then it follows from Theorem 2.3.7 and (2.3.12) that

$$\begin{aligned} (A^\sharp A)^{\frac{1}{2}} &= \left(|\pi(A)|, U_A^*\lambda(A), (U_A^*\lambda(A))^*\right) \\ &= \left(|\pi(B)|, U_B^*\lambda(B), (U_B^*\lambda(B))^*\right) \\ &= (B^\sharp B)^{\frac{1}{2}}. \end{aligned}$$

Suppose, conversely, that $\ker \pi(A) \neq \{0\}$. Then take a non-zero vector $x_0 \in \ker \pi(A) = \ker|\pi(A)|$, and define an element B of $T^*(\mathcal{H})$ by

$$B = \left(|\pi(A)|, U_A^*\lambda(A) + x_0, (U_A^*\lambda(A^\sharp) + x_0)^*\right).$$

Then we can prove that

$$\begin{aligned} B^\sharp B &= \left(\pi(A)^*\pi(A), |\pi(A)|U_A^*\lambda(A), (|\pi(A)|U_A^*\lambda(A^\sharp))^*\right) \\ &= \left(\pi(A)^*\pi(A), \pi(A)^*\lambda(A), (\pi(A)^*\lambda(A^\sharp))^*\right) \\ &= A^\sharp A, \end{aligned}$$

and that

$$\begin{aligned} (B^\sharp B)^{\frac{1}{2}} &= \left(|\pi(A)|, U_A^*\lambda(A) + x_0, (U_A^*\lambda(A) + x_0)^*\right) \\ &\neq (A^\sharp A)^{\frac{1}{2}}. \end{aligned}$$

Thus the square root $(A^\sharp A)^{\frac{1}{2}}$ of $A^\sharp A$ is not unique. This completes the proof.

Corollary 2.3.9 *Let $A = A^\sharp \in T^*(\mathcal{H})$. If the square root $(A^\sharp A)^{\frac{1}{2}}$ of $A^\sharp A$ is unique, then A is semisimple.*

Proof By Proposition 2.3.8 we have

$$\begin{aligned}(A^\sharp A)^{\frac{1}{2}} \text{ is unique if and only if } &\ker \pi(A) = \{0\} \\ \text{if and only if } &P_A = \text{Proj}\,[R(\pi(A))] = I\end{aligned}$$

which implies by Corollary 2.3.6 that

$$A = A_s, \quad \text{namely, } A \text{ is semisimple.}$$

We next discuss the polar decomposition of $A \in T^*(\mathcal{H})$. For $A \in T^*(\mathcal{H})$ and $B_0 \in B(\mathcal{H})$ we define an element $B_0 A$ of $T^*(\mathcal{H})$ by

$$B_0 A = \left(B_0 \pi(A), B_0 \lambda(A), (B_0 \lambda(A^\sharp))^*\right).$$

Proposition 2.3.10 *Let $A \in T^*(\mathcal{H})$. Then the following statements hold.*

(1) Suppose that $\lambda(A) \in [R(\pi(A))]$ and $\lambda(A^\sharp) \in [R(\pi(A)^*)]$. Then

$$A + A^\sharp = U_A (A^\sharp A)^{\frac{1}{2}} + U_A^* (A A^\sharp)^{\frac{1}{2}}.$$

(2) Suppose that $A = A^\sharp$, then

$$A \text{ is semisimple if and only if } A = U_A (A^\sharp A)^{\frac{1}{2}}.$$

In this case, this decomposition is unique, in the sense that if $A = U_B (B^\sharp B)^{\frac{1}{2}}$ for some $B \in T^*(\mathcal{H})$, then $U_B = U_A$ and $(B^\sharp B)^{\frac{1}{2}} = (A^\sharp A)^{\frac{1}{2}}$.

Proof

(1) Since $U_A U_A^* = \text{Proj}\,[R(\pi(A))]$ and $U_A^* U_A = \text{Proj}\,[R(\pi(A)^*)]$, it follows that

$$\begin{aligned}U_A (A^\sharp A)^{\frac{1}{2}} &= \left(U_A |\pi(A)|, U_A U_A^* \lambda(A), (U_A U_A^* \lambda(A))^*\right) \\ &= (\pi(A), \lambda(A), \lambda(A)^*)\end{aligned}$$

and

$$\begin{aligned}U_A^* (A A^\sharp)^{\frac{1}{2}} &= \left(U_A^* |\pi(A)^*|, U_A^* U_A \lambda(A^\sharp), (U_A^* U_A \lambda(A^\sharp))^*\right) \\ &= (\pi(A)^*, \lambda(A^\sharp), \lambda(A^\sharp)^*).\end{aligned}$$

Therefore

$$A + A^{\sharp} = U_A(A^{\sharp}A)^{\frac{1}{2}} + U_A^*(AA^{\sharp})^{\frac{1}{2}}. \tag{2.3.13}$$

(2) Suppose that $A = A^{\sharp}$. Then, $U_A = U_A^*$, and so by (2.3.13) we obtain

$$A = U_A(A^{\sharp}A)^{\frac{1}{2}}. \tag{2.3.14}$$

Conversely, suppose that $A = U_A(A^{\sharp}A)^{\frac{1}{2}}$. Then we have

$$\lambda(A) = U_A U_A^* \lambda(A) = P_A \lambda(A),$$

which implies that $A = A_s$. Suppose that $A = U_A(A^{\sharp}A)^{\frac{1}{2}} = U_B(B^{\sharp}B)^{\frac{1}{2}}$ for some $B \in T^*(\mathcal{H})$. Then we have

$$\pi(A) = \pi(B) \quad \text{and} \quad \lambda(A) = U_B U_B^* \lambda(B),$$

which yields that

$$U_A = U_B = U_B^* \quad \text{and} \quad \lambda(A) = P_A \lambda(B). \tag{2.3.15}$$

Furthermore, since

$$\begin{aligned}\|U_A(I - P_A)\lambda(B)\|^2 &= \left(U_A^2(I - P_A)\lambda(B) | (I - P_A)\lambda(B)\right)\\ &= 0,\end{aligned}$$

we have $U_A(I - P_A)\lambda(B) = 0$. Hence, it follows from (2.3.15) that

$$\begin{aligned}U_A\lambda(B) &= U_A P_A \lambda(B) + U_A(I - P_A)\lambda(B)\\ &= U_A\lambda(A),\end{aligned}$$

which implies that

$$\begin{aligned}(B^{\sharp}B)^{\frac{1}{2}} &= \left(|\pi(B)|, U_B^*\lambda(B), (U_B^*\lambda(B^{\sharp}))^*\right)\\ &= \left(|\pi(A)|, U_A\lambda(A), (U_A\lambda(A))^*\right)\\ &= (A^{\sharp}A)^{\frac{1}{2}}.\end{aligned}$$

Thus, the decomposition of A in (2.3.14) is unique; hence A is semisimple by Corollary 2.3.9. This completes the proof.

Finally, we consider a functional calculus of a general $A \in T^*(\mathcal{H})$. For any polynomial: $p(t) = \alpha_0 + \alpha_1 t + \cdots + \alpha_n t^n$, we can check that

$$p(AA^\sharp) \notin T^*(\mathcal{H}) \quad \text{if} \quad \alpha_0 \neq 0,$$

and that

$$\begin{aligned} p(AA^\sharp)A &= \alpha_0 A + \alpha_1 (AA^\sharp)A + \cdots + \alpha_n (AA^\sharp)^n A \\ &= \big(p(\pi(A)\pi(A)^*)\pi(A),\, p(\pi(A)\pi(A)^*)\lambda(A),\, (\bar{p}(\pi(A)^*\pi(A))\lambda(A^\sharp))^*\big) \\ &\in T^*(\mathcal{H}). \end{aligned}$$

Hence, for a general function f it is natural to write

$$f(AA^\sharp)A = \big(f(\pi(A)\pi(A)^*)\pi(A),\, f(\pi(A)\pi(A)^*)\lambda(A),\, (\bar{f}(\pi(A)^*\pi(A))\lambda(A^\sharp))^*\big).$$

We consider that $f(AA^\sharp)A$ belongs to $T^*(\mathcal{H})$ for what kind of functions f. Let Ω be a compact space and $C(\Omega)$ the set of all complex-valued continuous functions on Ω. Then $C(\Omega)$ is a commutative C^*-algebra equipped with the usual function operations $(f + g, \alpha f, fg)$, the involution $f \to f^* = \bar{f}$ and the uniform norm $\|f\|_u := \sup_{t\in\Omega} |f(t)|$. We now construct a simple functional calculus of non-self adjoint trio observables which will be used in the proof of Theorem 2.3.12:

Theorem 2.3.11 *Let $A \in T^*(\mathcal{H})$ and $f \in C[0, \|\pi(A)\|^2]$. Then $f(AA^\sharp)A$ belongs to the closed subspace $\mathfrak{M}(A)$ of $CT^*(A)$ generated by $\{(AA^\sharp)^n A;\ n = 0, 1, \ldots\}$, and $f \in C[0, \|\pi(A)\|^2] \to f(AA^\sharp)A \in \mathfrak{M}(A)$ is a continuous linear map.*

Proof It is clear that $f(A^\sharp A)A \in T^*(\mathcal{H})$ for any $f \in C[0, \|\pi(A)\|^2]$, and in particular it belongs to $\mathfrak{M}(A)$ for any polynomial. Hence it follows from the Weierstrass approximation theorem that $f(AA^\sharp)A \in \mathfrak{M}(A)$ for any $f \in C[0, \|\pi(A)\|^2]$. Furthermore, it is clear that $f \to f(AA^\sharp)A$ is a continuous linear map of the Banach space $C[0, \|\pi(A)\|^2]$ into the Banach space $\mathfrak{M}(A)$. This completes the proof.

In Corollary 2.3.6 we have shown that every *self-adjoint* element A of $T^*(\mathcal{H})$ is decomposed into the semisimple part $A_s \in CT^*(A)$ and the nilpotent part $A_n \in CT^*(A)$. Using Theorem 2.3.11, we show that this decomposition is possible for a general element of $T^*(\mathcal{H})$. For $A \in T^*(\mathcal{H})$, write

$$\begin{aligned} P(CT^*(A)) &= \text{the } CT^*\text{- algebra generated by } \{ST;\ S, T \in CT^*(A)\}, \\ N(CT^*(A)) &= \{T \in CT^*(A);\ \pi(T) = 0\}. \end{aligned}$$

We now define elements A_s and A_n of $T^*(\mathcal{H})$ as follows:

$$A_s = \left(\pi(A), P_A\lambda(A), (P_{A^\sharp}\lambda(A^\sharp))^*\right),$$
$$A_n = \left(0, (I - P_A)\lambda(A), ((I - P_{A^\sharp})\lambda(A^\sharp))^*\right).$$

Theorem 2.3.12 *Every element A of $T^*(\mathcal{H})$ is decomposed into:*

$$A = A_s + A_n,$$

where $A_s \in P(CT^*(A))$ and $A_n \in N(CT^*(A))$.

Proof Put

$$\chi_0(t) = \begin{cases} 1, \ t > 0 \\ 0, \ t = 0 \end{cases}$$

and

$$\begin{aligned}&\chi_0(AA^\sharp)A \\ &= \left(\chi_0(\pi(A)\pi(A)^*)\pi(A), \chi_0(\pi(A)\pi(A)^*)\lambda(A), (\chi_0(\pi(A)^*\pi(A))\lambda(A^\sharp))^*\right).\end{aligned}$$

Let $|\pi(A)^*| = \int_0^{\|\pi(A)\|} t dE(t)$ be the spectral resolution of $|\pi(A)^*|$. The calculation:

$$\begin{aligned}\chi_0(\pi(A)\pi(A)^*) &= \int_0^{\|\pi(A)\|} \chi_0(t^2) dE(t) \\ &= I - E(0) \\ &= I - \text{Proj } \ker|\pi(A)^*| \\ &= I - \text{Proj } R(\pi(A))^\perp \\ &= P_A, \end{aligned} \tag{2.3.16}$$

implies that

$$\begin{aligned}\chi_0(AA^\sharp)A &= \left(P_A\pi(A), P_A\lambda(A), (P_{A^\sharp}\lambda(A^\sharp))^*\right) \\ &= \left(\pi(A), P_A\lambda(A), (P_{A^\sharp}\lambda(A^\sharp))^*\right) \\ &= A_s. \end{aligned} \tag{2.3.17}$$

Now, consider a sequence $\{f_n\}$ of continuous functions on $\mathbb{R}_+$:

$$f_n(t) = \begin{cases} 1, & t > \frac{1}{n} \\ nt, & 0 \leqq t \leqq \frac{1}{n}. \end{cases}$$

By Theorem 2.3.11 we have

$$f_n(AA^\sharp)A \in \mathfrak{M}(A) \subset P(CT^*(A)). \tag{2.3.18}$$

Since

$$\sup_{t\in\mathbb{R}_+} |f_n(t)\sqrt{t} - \chi_0(t)\sqrt{t}| \leqq \frac{1}{n}, \quad n \in \mathbb{N},$$

it follows from (2.3.16) that

$$\begin{aligned} &\lim_{n\to\infty} \|f_n(\pi(A)\pi(A)^*)\pi(A) - \chi_0(\pi(A)\pi(A)^*)\pi(A)\| \\ &= \lim_{n\to\infty} \|f_n(\pi(A)\pi(A)^*)\pi(A) - P_A\pi(A)\| \\ &= 0. \end{aligned}$$

Furthermore, since $\lim_{t\to\infty} f_n(t) = \chi_0(t)$, $t \in \mathbb{R}_+$, it follows that

$$\begin{aligned} &\lim_{n\to\infty} \|f_n(\pi(A)\pi(A)^*)\lambda(A) - \chi_0(\pi(A)\pi(A)^*)\lambda(A)\| \\ &= \lim_{n\to\infty} \|f_n(\pi(A)\pi(A)^*)\lambda(A) - P_A\lambda(A)\| \\ &= 0, \end{aligned}$$

and that

$$\lim_{n\to\infty} \|f_n(\pi(A)^*\pi(A))\lambda(A^\sharp) - P_{A^\sharp}\lambda(A^\sharp)\| = 0,$$

which implies by (2.3.17) that

$$\lim_{n\to\infty} \|f_n(AA^\sharp)A - A_s\| = 0.$$

Hence it follows from (2.3.18) that $A_s \in P(CT^*(A))$, so that $A_n = A - A_s \in N(CT^*(A))$. This completes the proof.

The element A_s of $P(CT^*(A))$ in Theorem 2.3.12 is called the *semisimple part* of A, and the element A_n of $N(CT^*(A))$ is called the *nilpotent part* of A. If $A = A_s$, then A is called *semisimple*, and if $A = A_n$, then it is called *nilpotent*.

Notes For the spectrum, the spectral resolution, the functional calculus and the polar decomposition of a bounded linear operator in Hilbert space used in this section, refer to Appendix A.

2.4 $*$-Automorphisms of Observable Algebras

To investigate $*$-automorphisms of $T^*(\mathcal{H})$, we first define three fundamental $*$-automorphisms of $T^*(\mathcal{H})$. For any unitary operator U on $\mathcal{H}$, $\alpha \in \mathbb{C}$ and $g \in \mathcal{H}$ we put

$$\begin{aligned} I_U(A) &= (U\pi(A)U^*, U\lambda(A), (U\lambda(A^\sharp))^*), \\ S_\alpha(A) &= (\pi(A), \alpha\lambda(A), (\alpha\lambda(A^\sharp))^*), \\ T_g(A) &= (\pi(A), \pi(A)g + \lambda(A), (\pi(A)^*g + \lambda(A^\sharp))^*) \end{aligned}$$

for $A \in T^*(\mathcal{H})$. Then it is easily shown that I_U, S_α and T_g are continuous $*$-automorphisms of the CT^*-algebra $T^*(\mathcal{H})$. They are called the *unitary transform*, the *similar transform* and the *translation* of $T^*(\mathcal{H})$, respectively. The following theorem is the main result of this section:

Theorem 2.4.1 *For any $*$-automorphism σ of the CT^*-algebra $T^*(\mathcal{H})$ there exist a unitary operator U on $\mathcal{H}$, $\alpha \in \mathbb{C}$ and $g \in \mathcal{H}$ such that*

$$\sigma = T_g \circ I_U \circ S_\alpha.$$

Proof Put

$$\sigma_0(\pi(A)) = \pi(\sigma(A)), \quad A \in T^*(\mathcal{H}).$$

Suppose $\pi(A) = 0$. Then, since $A^\sharp A = 0$, we have $\sigma(A)^\sharp\sigma(A) = \sigma(A^\sharp A) = 0$, which implies that $\pi(\sigma(A)) = 0$. Hence σ_0 is well-defined, and it is easily shown that σ_0 is a $*$-automorphism of $B(\mathcal{H})$. Then there exists a unitary operator U on $\mathcal{H}$ such that

$$\sigma_0(\pi(A)) = U\pi(A)U^*, \quad A \in T^*(\mathcal{H}) \tag{2.4.1}$$

(see Notes of this section). Here, for any $X_0 \in B(\mathcal{H})$ and $x \in \mathcal{H}$ we define elements $\tilde{X}_0$ and $\tilde{x}$ of $T^*(\mathcal{H})$ by

$$\begin{aligned} \tilde{X}_0 &= (X_0, 0, 0), \\ \tilde{x} &= (0, x, 0). \end{aligned}$$

By (2.4.1) there exist g and h of $\mathcal{H}$ such that

$$\begin{aligned}\sigma(\tilde{I}) &= (UIU^*, g, h^*)\\ &= (I, g, h^*).\end{aligned}$$

Since $\tilde{I}^\sharp = \tilde{I}$, we have

$$\sigma(\tilde{I}) = \sigma(\tilde{I}^\sharp) = (I, h, g^*),$$

hence, $g = h$ and

$$\sigma(\tilde{I}) = (I, g, g^*). \tag{2.4.2}$$

By (2.4.1) again, there exist element ξ and η of $\mathcal{H}$ such that

$$\sigma(\tilde{X}_0) = (UX_0U^*, \xi, \eta^*).$$

Since $\tilde{X}_0\tilde{I} = \tilde{X}_0$ and $\tilde{I}\tilde{X}_0 = \tilde{X}_0$, we have

$$\begin{aligned}\sigma(\tilde{X}_0) &= \sigma(\tilde{X}_0\tilde{I}) = \sigma(\tilde{X}_0)\sigma(\tilde{I})\\ &= (UX_0U^*, UX_0U^*g, \eta^*)\end{aligned}$$

and

$$\begin{aligned}\sigma(\tilde{X}_0) &= \sigma(\tilde{I})\sigma(\tilde{X}_0)\\ &= (UX_0U^*, \xi, (UX_0^*U^*g)^*),\end{aligned}$$

which implies by (2.4.2) that for $X_0 \in B(\mathcal{H})$,

$$\sigma(\tilde{X}_0) = (UX_0U^*, UX_0U^*g, (UX_0U^*g)^*) \tag{2.4.3}$$

and

$$(T_{-g} \circ \sigma)(\tilde{X}_0) = (UX_0U^*, 0, 0). \tag{2.4.4}$$

Now, put

$$\sigma' = I_{U^*} \circ T_{-g} \circ \sigma.$$

By (2.4.4) we obtain

$$\sigma'(\tilde{X}_0) = \tilde{X}_0, \quad X_0 \in B(\mathcal{H}). \tag{2.4.5}$$

Let $x \in \mathcal{H}$. Since $\sigma(\tilde{x}^\sharp)\sigma(\tilde{x}) = \sigma(\tilde{x}^\sharp\tilde{x}) = O$, there exist ξ_x and η_x of $\mathcal{H}$ such that

$$\sigma'(\tilde{x}) = (0, \xi_x, \eta_x^*),$$

so that by (2.4.2),

$$\sigma'(\tilde{x}) = \sigma'(\tilde{I})\sigma'(\tilde{x}) = (0, \xi_x, 0). \tag{2.4.6}$$

We here define a linear map Λ of $\mathcal{H}$ onto $\mathcal{H}$ by

$$\Lambda x = \xi_x, \quad x \in \mathcal{H}.$$

Then it follows from (2.4.5) and (2.4.6) that

$$\begin{aligned}(0, \Lambda X_0 x, 0) &= \sigma'((0, X_0 x, 0)) = \sigma'(\tilde{X}_0\tilde{x}) \\ &= \sigma'(\tilde{X}_0)\sigma'(\tilde{x}) = \tilde{X}_0\sigma'(\tilde{x}) \\ &= (0, X_0\Lambda x, 0)\end{aligned}$$

for all $X_0 \in B(\mathcal{H})$ and $x \in \mathcal{H}$, which implies that $\Lambda X_0 = X_0\Lambda$ for all $X_0 \in B(\mathcal{H})$. Though we don't know whether Λ is bounded or not, we can show that $\Lambda = \alpha I$ for some $\alpha \in \mathbb{C}$. Indeed, since

$$\begin{aligned}(\Lambda x|\eta)\xi &= (\xi \otimes \bar{\eta})\Lambda x \\ &= \Lambda(\xi \otimes \bar{\eta})x \\ &= (x|\eta)\Lambda\xi\end{aligned}$$

for all $x, \xi, \eta \in \mathcal{H}$, where $(\xi \otimes \bar{\eta})x := (x|\eta)\xi$, it follows that

$$(\Lambda x|\eta) = (\Lambda\xi|\xi)(x|\eta)$$

for all $x, \eta \in \mathcal{H}$ and $\xi \in \mathcal{H}_1 := \{\zeta \in \mathcal{H}; \|\zeta\| = 1\}$. Furthermore, we can prove that $(\Lambda\xi|\xi)$ does not depend on the method of taking $\xi \in \mathcal{H}_1$. Hence, $\Lambda = \alpha I$ for some $\alpha \in \mathbb{C}$, so that

$$\sigma'((0, x, 0)) = (0, \alpha x, 0)$$

and

$$\begin{aligned}\sigma'((0, 0, y^*)) &= \sigma'((0, y, 0)^\sharp) \\ &= \sigma'((0, y, 0))^\sharp \\ &= (0, \alpha y, 0)^\sharp \\ &= (0, 0, (\alpha y)^*)\end{aligned}$$

for all $x,\ y \in \mathcal{H}$. Thus we have

$$\sigma'((0, x, y^*)) = (0, \alpha x, (\alpha y)^*).$$

Therefore, it follows from (2.4.5) that

$$\begin{aligned}\sigma'(X) &= \tilde{X}_0 + (0, \alpha x, (\alpha y)^*)\\ &= (X_0, \alpha x, (\alpha y)^*)\\ &= S_\alpha(X)\end{aligned}$$

for all $X \in T^*(\mathcal{H})$, which yields that

$$\sigma = T_g \circ I_U \circ S_\alpha.$$

This completes the proof.

By Theorem 2.4.1 every ∗-automorphism of the CT^*-algebra $T^*(\mathcal{H})$ always continuous.

We next discuss ∗-automorphisms of $Q^*(\mathcal{H})$. Put

$$\tau(A) = (\pi(A), \lambda(A), \lambda(A^\sharp)^*)$$

for $A = (\pi(A), \lambda(A), \lambda(A^\sharp)^*, \mu(A)) \in Q^*(\mathcal{H})$. Then, $\tau(Q^*(\mathcal{H})) = T^*(\mathcal{H})$. For any unitary operator U on $\mathcal{H}$, $\alpha \in \mathbb{C}$ and $g \in \mathcal{H}$ we define continuous ∗-automorphisms $I_U^q(A)$, $S_\alpha^q(A)$ and $T_g^q(A)$ of the Banach ∗-algebra $Q^*(\mathcal{H})$ by

$$\begin{aligned}I_U^q(A) &= (I_U(\tau(A)), \mu(A)),\\ S_\alpha^q(A) &= (S_\alpha(\tau(A)), |\alpha|^2\mu(A)),\\ T_g^q(A) &= \big(T_g(\tau(A)), (\mu(A) + (\lambda(A)|g) + (g|\lambda(A^\sharp)) + (\pi(A)g|g))\big)\end{aligned}$$

for $A \in Q^*(\mathcal{H})$, which are called the *unitary transform*, the *similar transform* and the *translation* of $Q^*(\mathcal{H})$, respectively.

Lemma 2.4.2 *Suppose that σ is a ∗-automorphism of the CQ^*-algebra $Q^*(\mathcal{H})$. Then there exist a unitary operator U on $\mathcal{H}$, $\alpha \in \mathbb{C}$ and $g \in \mathcal{H}$ such that*

$$\sigma(AB) = (T_g^q \circ I_U^q \circ S_\alpha^q)(AB)$$

for all $A,\ B \in Q^*(\mathcal{H})$.

Proof Suppose $\tau(A) = O$. Then, $A = (0, 0, 0, \mu(A))$ and $A^\sharp = (0, 0, 0, \bar{\mu}(A))$. Let $\sigma(A) = (X_0, x, y, \alpha)$. Since $A^\sharp A = AA^\sharp = O$, we have

$$\begin{aligned} O &= \sigma(A^\sharp A) = \sigma(A)^\sharp \sigma(A) \\ &= (X_0^* X_0, X_0^* x, (X_0^* x)^*, \|x\|^2), \\ O &= \sigma(AA^\sharp) \\ &= (X_0 X_0^*, X_0 y, (X_0 y)^*, \|y\|^2), \end{aligned}$$

so that $X_0 = 0$, $x = y = 0$. Hence $\tau(\sigma(A)) = O$. Thus we can define a $*$-automorphism σ_τ of the CT^*-algebra $T^*(\mathcal{H})$ by $\sigma_\tau(\tau(A)) = \tau(\sigma(A))$, $A \in Q^*(\mathcal{H})$. By Theorem 2.4.1 there exist a unitary operator U on $\mathcal{H}$, $\alpha \in \mathbb{C}$ and $g \in \mathcal{H}$ such that

$$\sigma_\tau = T_g \circ I_U \circ S_\alpha, \tag{2.4.7}$$

so that

$$\begin{aligned} \pi(\sigma(A)) &= U\pi(A)U^*, \\ \lambda(\sigma(A)) &= U\pi(A)U^* g + \alpha U \lambda(A), \qquad (2.4.8) \\ \lambda(\sigma(A)^\sharp) &= U\pi(A)^* U^* g + \alpha U \lambda(A^\sharp) \qquad (2.4.9) \end{aligned}$$

for all $A \in Q^*(\mathcal{H})$. Hence it follows from (2.4.8) and (2.4.9) that

$$\begin{aligned} \mu(\sigma(AB)) &= \mu(\sigma(A)\sigma(B)) = (\lambda(\sigma(B))|\lambda(\sigma(A)^\sharp)) \\ &= (U\pi(B)U^* g + \alpha U \lambda(B)|U\pi(A)^* U^* g + \alpha U \lambda(A^\sharp)) \\ &= (U\pi(AB)U^* g|g) + (g|\alpha U \lambda(B^\sharp A^\sharp)) \\ &\quad + (\alpha U \lambda(AB)|g) + |\alpha|^2 \mu(AB). \end{aligned}$$

On the other hand, since

$$\begin{aligned} S_\alpha^q(AB) &= (S_\alpha(\tau(AB)), |\alpha|^2 \mu(AB)), \\ (I_U^q \circ S_\alpha^q)(AB) &= (I_U \circ S_\alpha(\tau(AB)), |\alpha|^2 \mu(AB)), \end{aligned}$$

we have

$$\begin{aligned} \mu((T_g^q \circ I_U^q \circ S_\alpha^q)(AB)) &= \mu((I_U^q \circ S_\alpha^q)(AB)) + (\lambda((I_U^q \circ S_\alpha^q)(AB)|g) \\ &\quad + (g|\lambda((I_U^q \circ S_\alpha^q)(AB)^\sharp)) + (\pi(I_U^q \circ S_\alpha^q)(AB)g|g) \\ &= \alpha^2 \mu(AB) + (\alpha U \lambda(AB)|g) + (g|\alpha U \lambda((AB)^\sharp)) \\ &\quad + (U\pi(AB)U^* g|g) \\ &= \mu(\sigma(AB)), \end{aligned}$$

which implies by (2.4.7) that

$$\begin{aligned}\sigma(AB) &= (\sigma_\tau(\tau(AB)), \mu(\sigma(AB))) \\ &= ((T_g \circ I_U \circ S_\alpha)\tau(AB), \mu((T_g^q \circ I_U^q \circ S_\alpha^q)(AB))) \\ &= (T_g^q \circ I_U^q \circ S_\alpha^q)(AB).\end{aligned}$$

This completes the proof.

By Lemma 2.4.2 we obtain the following result for $*$-automorphisms of the CQ^*-algebra $Q^*(\mathcal{H})$:

Theorem 2.4.3 *Suppose that σ is a $*$-automorphism of the CQ^*-algebra $Q^*(\mathcal{H})$. Then there exist a unitary operator U on $\mathcal{H}$, $\alpha \in \mathbb{C}$ and $g \in \mathcal{H}$ such that*

$$\sigma(A) = (T_g^q \circ I_U^q \circ S_\alpha^q)(A)$$

for all $A \in Q^*(\mathcal{H})$, that is,

$$\begin{aligned}\pi(\sigma(A)) &= U\pi(A)U^*, \\ \lambda(\sigma(A)) &= (U\pi(A)U^*)g + \alpha U\lambda(A), \\ \lambda^*(\sigma(A)) &= ((U\pi(A)^*U^*)g + \alpha U\lambda(A^\sharp))^*, \\ \mu(\sigma(A)) &= |\alpha|^2\mu(A) + (\alpha U\lambda(A)|g) + (g|\alpha U\lambda(A^\sharp)) \\ &\quad + (U\pi(A)U^*g|g).\end{aligned}$$

Proof Take an arbitrary $A \in Q^*(\mathcal{H})$. Then we define elements $\tilde{I}$, B, C of $Q^*(\mathcal{H})$ by

$$\begin{aligned}\tilde{I} &= (I, 0, 0, 0), \\ B &= \left(\frac{1}{2}\pi(A), \frac{1}{2}\lambda(A), \lambda(A^\sharp)^*, \frac{1}{2}\mu(A)\right), \\ C &= \left(\frac{1}{2}\pi(A), \lambda(A), \frac{1}{2}\lambda(A^\sharp)^*, \frac{1}{2}\mu(A)\right).\end{aligned}$$

Then, by the equality:

$$A = B\tilde{I} + \tilde{I}C + \mu(A)V,$$

we get

$$\sigma(A) = \sigma(B\tilde{I}) + \sigma(\tilde{I}C) + \mu(A)\sigma(V). \tag{2.4.10}$$

By Lemma 2.4.2, we obtain

$$\sigma(B\tilde{I}) = \left((T_g \circ I_U \circ S_\alpha)(\tau(B\tilde{I})), (g|\alpha U\lambda(A^\sharp)) + \frac{1}{2}(U\pi(A)U^* g|g)\right) \tag{2.4.11}$$

and

$$\sigma(\tilde{I}C) = \left((T_g \circ I_U \circ S_\alpha)(\tau(\tilde{I}B)), (\alpha U\lambda(A)|g) + \frac{1}{2}(U\pi(A)U^* g|g)\right). \tag{2.4.12}$$

For any $x \in \mathcal{H}$ with $\|x\| = 1$, define an element $\tilde{x}$ of $Q^*(\mathcal{H})$ by

$$\tilde{x} = (0, x, 0, 0).$$

Then, $\tilde{x}^\sharp = (0, 0, x^*, 0)$ and $\tilde{x}^\sharp\tilde{x} = (0, 0, 0, \|x\|^2) = V$. Hence, $\sigma(\tilde{x}^\sharp\tilde{x}) = \sigma(V)$. On the other hand, by Lemma 2.4.2

$$\begin{aligned}\sigma(\tilde{x}^\sharp\tilde{x}) &= (\tau(\sigma(\tilde{x}^\sharp\tilde{x})), |\alpha|^2) \\ &= |\alpha|^2 V.\end{aligned}$$

Thus,

$$\sigma(V) = |\alpha|^2 V. \tag{2.4.13}$$

By (2.4.10)–(2.4.13) we have

$$\begin{aligned}\tau(\sigma(A)) &= (T_g \circ I_U \circ S_\alpha)(\tau(A)), \\ \mu(\sigma(A)) &= (g|\alpha U\lambda(A^\sharp)) + (U\pi(A)U^* g|g) + |\alpha|^2\mu(A) \\ &= \mu((T_g^q \circ I_U^q \circ S_\alpha^q)(A)),\end{aligned}$$

which completes the proof.

By Theorem 2.4.3 every $*$-automorphism of $Q^*(\mathcal{H})$ is continuous.

Notes Every $*$-automorphism α of $B(\mathcal{H})$ is represented as

$$\alpha(A) = UAU^*, \quad A \in B(\mathcal{H})$$

for some unitary operator U on $\mathcal{H}$. Indeed, for any $\xi, \eta \in \mathcal{H}$ we denote by $\xi \otimes \bar{\eta}$ the one dimensional operator on $\mathcal{H}$ by

$$(\xi \otimes \bar{\eta})x = (x|\eta)\xi, \quad x \in \mathcal{H}.$$

Let $x_0 \in \mathcal{H}$ with $\|x_0\| = 1$. Then it is easily shown that $x_0 \otimes \bar{x}_0$ is a non-zero projection on $\mathcal{H}$. Hence, $\alpha(x_0 \otimes \bar{x}_0)$ is a non-zero projection on $\mathcal{H}$. Here, take an element ξ_0 of $\mathcal{H}$ such that $\|\xi_0\| = 1$ and $\alpha(x_0 \otimes \bar{x}_0)\xi_0 = \xi_0$. Then we have

$$(x_0 \otimes \bar{x}_0)A(x_0 \otimes \bar{x}_0) = (Ax_0|x_0)(x_0 \otimes \bar{x}_0)$$

for all $A \in B(\mathcal{H})$, which implies that

$$\begin{aligned}(\alpha(A)\xi_0|\xi_0) &= (\alpha((x_0 \otimes \bar{x}_0)A(x_0 \otimes \bar{x}_0))\xi_0|\xi_0) \\ &= (Ax_0|x_0).\end{aligned}$$

Furthermore, since $B(\mathcal{H})x = \alpha(B(\mathcal{H}))x = \mathcal{H}$ for any non-zero $x \in \mathcal{H}$, we can define a unitary operator U on $\mathcal{H}$ by

$$UAx_0 = \alpha(A)\xi_0.$$

Hence,

$$UAU^*\xi_0 = \alpha(A)\xi_0$$

for all $A \in B(\mathcal{H})$. Take an arbitrary $x \in \mathcal{H}$. Then there exists an element B of $B(\mathcal{H})$ such that $x = \alpha(B)\xi_0 = UBx_0$, which yields that

$$UAU^*x = UAB\xi_0 = \alpha(AB)\xi_0 = \alpha(A)x.$$

Thus, $\alpha(A) = UAU^*$ for all $A \in B(\mathcal{H})$.

For the spatiality of $*$-automorphisms of more general von Neumann algebras, refer to [35].

2.5 Locally Convex Topologies on T^*-algebras

As mentioned in Sect. 2.2, $T^*(\mathcal{H})$ is a Banach $*$-algebra with the norm: $\|A\| = \max(\|\pi(A)\|, \|\lambda(A)\|, \|\lambda(A^\sharp)\|)$, $A \in T^*(\mathcal{H})$. The topology defined by the norm $\|\cdot\|$ is called the *uniform topology* and is denoted by τ_u. We define other locally convex topologies on $T^*(\mathcal{H})$. The product topology on $T^*(\mathcal{H}) = B(\mathcal{H}) \times \mathcal{H} \times \mathcal{H}^*$ defined in terms of the weak (resp. the σ-weak) topology on $B(\mathcal{H})$ and the weak topology on $\mathcal{H}$ and $\mathcal{H}^*$ is called the *weak* (resp. the *σ-weak*) topology on $T^*(\mathcal{H})$ and is denoted by τ_w (resp. $\tau_{\sigma w}$). The product topology on $T^*(\mathcal{H})$ defined in terms of the strong (resp. the σ-strong, strong* and σ-strong*) topology on $B(\mathcal{H})$ (see Appendix C.3) and the norm-topology on $\mathcal{H}$ and $\mathcal{H}^*$ is called the *strong* (resp. the *σ-strong*, *strong** and *σ- strong**) topology on $T^*(\mathcal{H})$ and is denoted by τ_s (resp.

$\tau_{\sigma s}$, τ_s^* and $\tau_{\sigma s}^*$). To be concrete, the σ-strong* topology $\tau_{\sigma s}^*$ on $T^*(\mathcal{H})$ is defined by a family $\{p_x^*;\ \boldsymbol{x} = \{x_n\} \in \mathcal{H}_{\mathbb{N}}\}$ of seminorms:

$$p_x^*(A) = \left[\sum_{n=1}^{\infty} \|\pi(A)x_n\|^2 + \sum_{n=1}^{\infty} \|\pi(A)^* x_n\|^2 + \|\lambda(A)\|^2 + \|\lambda(A^\sharp)\|^2\right]^{\frac{1}{2}},$$

and the σ-weak topology $\tau_{\sigma w}$ is defined by a family $\{p_{\boldsymbol{x},\boldsymbol{y},F,G};\ \boldsymbol{x} = \{x_n\},\ \boldsymbol{y} = \{y_n\} \in \mathcal{H}_{\mathbb{N}},\ F,\ G \in \mathfrak{F}\}$ of seminorms: for $F = \{f_1, \ldots, f_l\}$, $G = \{g_1, \ldots, g_m\} \in \mathfrak{F}$,

$$p_{\boldsymbol{x},\boldsymbol{y},F,G}(A) = \left[|\sum_{n=1}^{\infty}(\pi(A)x_n|y_n)| + |\sum_{k=1}^{l}(\lambda(A)|f_k)| + |\sum_{j=1}^{m}(g_j|\lambda(A^\sharp)|\right],$$

where $\mathcal{H}_{\mathbb{N}} := \{\{x_n\} \subset \mathcal{H};\ \sum_{n=1}^{\infty} \|x_n\|^2 < \infty\}$, and $\mathfrak{F}$ is the set of all finite subsets of $\mathcal{H}$. Similarly, other topologies τ_s^*, τ_s, $\tau_{\sigma s}$ and τ_w are concretely defined by families of seminorms.

As for their topologies we have the following

$$\begin{array}{ccccc} \tau_w & \prec & \tau_s & \prec & \tau_{\sigma s} \\ & & \curlywedge & & \curlywedge \\ & & \tau_s^* & \prec & \tau_{\sigma s}^* \prec \tau_u, \end{array} \tag{2.5.1}$$

where $\tau_1 \prec \tau_2$ means that a topology τ_2 is finer than a topology τ_1.

Let τ be one of the above topologies on $T^*(\mathcal{H})$. We denote $T^*(\mathcal{H})$ equipped with τ by $T^*(\mathcal{H})[\tau]$, and denote the induced topology of $T^*(\mathcal{H})[\tau]$ on a subset $\mathfrak{A}$ of $T^*(\mathcal{H})$ by $\mathfrak{A}[\tau]$. We denote the closure of $\mathfrak{A}[\tau]$ in $T^*(\mathcal{H})$ by $\bar{\mathfrak{A}}[\tau]$. Then we can easily show the following

Proposition 2.5.2 *Let $\mathfrak{A}$ be a T^*-algebra on $\mathcal{H}$. Then $\mathfrak{A}[\tau_w]$, $\mathfrak{A}[\tau_{\sigma w}]$, $\mathfrak{A}[\tau_s^*]$ and $\mathfrak{A}[\tau_{\sigma s}^*]$ are locally convex $*$-algebras. In particular, $T^*(\mathcal{H})[\tau_s^*]$ and $T^*(\mathcal{H})[\tau_{\sigma s}^*]$ are complete locally convex $*$-algebras.*

For the definition of locally convex $(*)$-algebras refer to Appendix C.4. $\mathfrak{A}[\tau_s]$ and $\mathfrak{A}[\tau_{\sigma s}]$ are not necessarily locally convex algebras. Indeed, suppose that a net $\{A_\alpha\}$ in $\mathfrak{A}$ converges strongly to an element A of $\mathfrak{A}$. For an element B of $\mathfrak{A}$ we have $A_\alpha B \to AB$, strongly, but BA_α does not necessarily converges strongly to BA because $\pi(A_\alpha)^*\lambda(B)$ does not necessarily converge to $\pi(A)^*\lambda(B)$ under the norm topology on $\mathcal{H}$. Hence the multiplication of $\mathfrak{A}[\tau_s]$ is not separately continuous, which means that $\mathfrak{A}[\tau_s]$ is not a locally convex algebra in general. Similarly, $\mathfrak{A}[\tau_{\sigma s}]$ is not a locally convex algebra in general. We next consider the closures of a T^*-algebra $\mathfrak{A}$ for their topologies. Since $T^*(\mathcal{H})[\tau_s]$, $T^*(\mathcal{H})[\tau_{\sigma s}]$, $T^*(\mathcal{H})[\tau_s^*]$ and $T^*(\mathcal{H})[\tau_{\sigma s}^*]$ are complete, the closures $\bar{\mathfrak{A}}[\tau_s]$, $\bar{\mathfrak{A}}[\tau_{\sigma s}]$, $\bar{\mathfrak{A}}[\tau_s^*]$ and $\bar{\mathfrak{A}}[\tau_{\sigma s}^*]$ of $\mathfrak{A}$ under the strong, σ-strong, strong* and σ-strong* topologies, respectively are contained in $T^*(\mathcal{H})$, but the completions of $\mathfrak{A}$ under the weak topology and the σ-weak topology are not necessarily contained in $T^*(\mathcal{H})$, so that we denote by $\bar{\mathfrak{A}}[\tau_w]$ and $\bar{\mathfrak{A}}[\tau_{\sigma w}]$ the closure of $\mathfrak{A}$ in $T^*(\mathcal{H})$ under the weak and the σ-weak topologies, respectively. Now we have the following

Proposition 2.5.3 *Let $\mathfrak{A}$ be a T^*-algebra on $\mathcal{H}$. Then, all of the closures of $\mathfrak{A}$ under the topologies τ_w, $\tau_{\sigma w}$, τ_s, $\tau_{\sigma s}$, τ_s^* and $\tau_{\sigma s}^*$ coincide.*

Proof By (2.5.1) we have

$$\begin{array}{ccccccc} \bar{\mathfrak{A}}[\tau_{\sigma s}] & \subset & \bar{\mathfrak{A}}[\tau_s] & \subset & \bar{\mathfrak{A}}[\tau_{\sigma w}] & \subset & \bar{\mathfrak{A}}[\tau_w] \\ \cup & & \cup & & & & \\ \bar{\mathfrak{A}}[\tau_{\sigma s}^*] & \subset & \bar{\mathfrak{A}}[\tau_s^*] & & & & \end{array} \tag{2.5.2}$$

and by the von Neumann type density theorem [Theorem 3.1.1] in Sect. 3.1

$$\bar{\mathfrak{A}}[\tau_{\sigma s}^*] = \bar{\mathfrak{A}}[\tau_w].$$

Hence, all of their closures of $\mathfrak{A}$ coincide. This completes the proof.

2.6 Commutants and Bicommutants of T^*-algebras

In this section we define and investigate three commutants $\mathfrak{A}^\pi$, $\mathfrak{A}^\tau$ and $\mathfrak{A}^\rho$, and three bicommutants $\mathfrak{A}^{\pi\pi}$, $\mathfrak{A}^{\tau\tau}$ and $\mathfrak{A}^{\rho\rho}$ of a T^*-algebra $\mathfrak{A}$. We first define commutants of T^*-algebras.

Definition 2.6.1 Let $\mathfrak{A}$ be a T^*-algebra on $\mathcal{H}$. We define the commutants $\mathfrak{A}^\pi$, $\mathfrak{A}^\tau$ and $\mathfrak{A}^\rho$ of $\mathfrak{A}$ as follows:

$$\begin{aligned} \mathfrak{A}^\pi &= \{K \in T^*(\mathcal{H});\quad \pi(K) \in \pi(\mathfrak{A})' \text{ (the commutant of } \pi(\mathfrak{A}))\}, \\ \mathfrak{A}^\tau &= \{K \in T^*(\mathcal{H});\quad AK = KA \text{ for all } A \in \mathfrak{A}\}, \\ \mathfrak{A}^\rho &= \{K \in \mathfrak{A}^\tau;\quad (\lambda(A)|\lambda(K^\sharp)) = (\lambda(K)|\lambda(A^\sharp)) \text{ for all } A \in \mathfrak{A}\}. \end{aligned}$$

Then we have the following

Proposition 2.6.2 *The commutants $\mathfrak{A}^\pi$, $\mathfrak{A}^\tau$ and $\mathfrak{A}^\rho$ of $\mathfrak{A}$ are CT^*-algebras which are closed under all of τ_w, $\tau_{\sigma w}$, τ_s, $\tau_{\sigma s}$, τ_s^* and $\tau_{\sigma s}^*$ topologies and*

$$(\mathfrak{A}^\tau)^2 \subset \mathfrak{A}^\rho \subset \mathfrak{A}^\tau \subset \mathfrak{A}^\pi. \tag{2.6.1}$$

The commutants $\mathfrak{A}^\rho$, $\mathfrak{A}^\tau$ and $\mathfrak{A}^\pi$ don't coincide in general.

Proof It is easily shown that the commutants $\mathfrak{A}^\pi$, $\mathfrak{A}^\tau$ and $\mathfrak{A}^\rho$ of $\mathfrak{A}$ are CT^*-algebras satisfying (2.6.1). Suppose now that a net $\{K_\alpha\}$ in $\mathfrak{A}^\tau$ converges weakly to an element K of $T^*(\mathcal{H})$. Then, for any $A \in \mathfrak{A}$, $K_\alpha A \underset{\alpha}{\to} KA$, weakly, that is,

$$\pi(K_\alpha)\pi(A) \underset{\alpha}{\to} \pi(K)\pi(A), \quad \text{weakly},$$
$$\pi(K_\alpha)\lambda(A) \underset{\alpha}{\to} \pi(K)\lambda(A), \quad \text{weakly},$$
$$\pi(A)^*\lambda(K_\alpha^\sharp) \underset{\alpha}{\to} \pi(A)^*\lambda(K^\sharp), \quad \text{weakly}$$

and similarly, $AK_\alpha \underset{\alpha}{\to} AK$, weakly. Because $K_\alpha A = AK_\alpha$, we get $KA = AK$, and so $K \in \mathfrak{A}^\tau$. Thus, $\mathfrak{A}^\tau$ is weakly closed. It is similarly shown that $\mathfrak{A}^\pi$ and $\mathfrak{A}^\rho$ are also weakly closed. In Example 2.6.4 later, it will see that each of their commutants does not coincide in general. This completes the proof.

As for the relationship between the fundamental $*$-automorphisms of $T^*(\mathcal{H})$ and the commutants of CT^*-algebras, we get the following

Proposition 2.6.3 *Let $\mathfrak{A}$ be a CT^*-algebra on $\mathcal{H}$ and σ a fundamental $*$-automorphism of $T^*(\mathcal{H})$. Then,*

$$\sigma(\mathfrak{A})^\pi = \sigma(\mathfrak{A}^\pi),\ \sigma(\mathfrak{A})^\tau = \sigma(\mathfrak{A}^\tau) \text{ and } \sigma(\mathfrak{A})^\rho = \sigma(\mathfrak{A}^\rho).$$

Proof Let $\sigma = T_g$, $g \in \mathcal{H}$. The equality:

$$T_g(A) = (\pi(A), \pi(A)g + \lambda(A), ((\pi(A))^*g + \lambda(A^\sharp))^*),\ \ A \in \mathfrak{A}$$

implies that

$$T_g(\mathfrak{A})^\pi = T_g(\mathfrak{A}^\pi). \tag{2.6.2}$$

Take an arbitrary $K \in T_g(\mathfrak{A})^\tau$. Then we can calculate that

$$\pi(K)\pi(A) = \pi(A)\pi(K),$$
$$\pi(K)(\pi(A)g + \lambda(A)) = \pi(A)\lambda(K),$$
$$\pi(K)^*(\pi(A)^*g + \lambda(A^\sharp)) = \pi(A)^*\lambda(K^\sharp)$$

for all $A \in \mathfrak{A}$, so that

$$\pi(K)\lambda(A) = \pi(A)(\lambda(K) - \pi(K)g),$$
$$\pi(K)^*\lambda(A^\sharp) = \pi(A)^*(\lambda(K^\sharp) - \pi(K)^*g).$$

Therefore, $X := T_{-g}(K) \in \mathfrak{A}^\tau$ and $K = T_g(X) \in T_g(\mathfrak{A}^\tau)$. Thus,

$$T_g(\mathfrak{A})^\tau \subset T_g(\mathfrak{A}^\tau). \tag{2.6.3}$$

Conversely suppose $K \in T_g(\mathfrak{A})^\rho$. Since $T_g(\mathfrak{A})^\rho \subset T_g(\mathfrak{A})^\tau \subset T_g(\mathfrak{A}^\tau)$ by (2.6.3), there exists an element X of $\mathfrak{A}^\tau$ such that $K = T_g(X)$, which implies that

$$\begin{aligned}(\lambda(K)|\lambda(T_g(A)^\sharp)) &= (\lambda(X) + \pi(X)g|\pi(A)^*g + \lambda(A^\sharp)) \\ &= (\lambda(X)|\pi(A)^*g) + (\lambda(X)|\lambda(A^\sharp)) \\ &\quad + (\pi(X)g|\pi(A)^*g) + (\pi(X)g|\lambda(A^\sharp)) \\ &= (\lambda(A)|\pi(X)^*g) + (\lambda(X)|\lambda(A^\sharp)) \\ &\quad + (\pi(A)g|\pi(X)^*g) + (\pi(A)g|\lambda(X^\sharp))\end{aligned}$$

and that

$$\begin{aligned}(\lambda(T_g(A))|\lambda(K^\sharp)) &= (\pi(A)g|\lambda(X^\sharp)) + (\pi(A)g|\pi(X)^*g) \\ &\quad + (\lambda(A)|\lambda(X^\sharp)) + (\lambda(A)|\pi(X)^*g)\end{aligned}$$

for all $A \in \mathfrak{A}$. Hence we obtain

$$(\lambda(X)|\lambda(A^\sharp)) = (\lambda(A)|\lambda(X^\sharp))$$

for all $A \in \mathfrak{A}$. Therefore, $X \in \mathfrak{A}^\rho$ and $K = T_g(X) \in T_g(\mathfrak{A}^\rho)$. Thus,

$$T_g(\mathfrak{A})^\rho \subset T_g(\mathfrak{A}^\rho). \tag{2.6.4}$$

The converses of (2.6.3) and (2.6.4):

$$T_g(\mathfrak{A}^\tau) \subset T_g(\mathfrak{A})^\tau \quad \text{and} \quad T_g(\mathfrak{A}^\rho) \subset T_g(\mathfrak{A})^\rho$$

are easily shown. The cases of $\sigma = I_U$ and $\sigma = S_\alpha$ are proved in the same way as the case $\sigma = T_g$. This completes the proof.

We next we discuss the bicommutants $\mathfrak{A}^{\pi\pi} := (\mathfrak{A}^\pi)^\pi$, $\mathfrak{A}^{\tau\tau} := (\mathfrak{A}^\tau)^\tau$ and $\mathfrak{A}^{\rho\rho} := (\mathfrak{A}^\rho)^\rho$ of $\mathfrak{A}$. It is immediate that

$$\begin{aligned}\mathfrak{A}^{\pi\pi} &= \{A \in T^*(\mathcal{H});\quad \pi(A) \in \pi(\mathfrak{A})''\}, \\ \mathfrak{A} \subset \mathfrak{A}^T &:= \mathfrak{A}^{\pi\pi} \cap \mathfrak{A}^{\tau\tau} \cap \mathfrak{A}^{\rho\rho},\end{aligned}$$

but the bicommutants $\mathfrak{A}^{\pi\pi}$, $\mathfrak{A}^{\tau\tau}$ and $\mathfrak{A}^{\rho\rho}$ don't have mutual relations in general as seen in next Example 2.6.4. In Sect. 3.1, we will show that $\overline{\mathfrak{A}}[\tau^*_{\sigma s}] = \mathfrak{A}^T$ for every CT^*-algebra $\mathfrak{A}$ [the von Neumann type density theorem]. We here give commutants and bicommutants of special T^*-algebras.

Example 2.6.4

(1) Let $T^*(\mathcal{H}) = (B(\mathcal{H}), \mathcal{H}, \mathcal{H}^*)$. Then

$$T^*(\mathcal{H})^\pi = (\mathbb{C}I, \mathcal{H}, \mathcal{H}^*), \quad T^*(\mathcal{H})^\tau = T^*(\mathcal{H})^\rho = \{O\},$$

and

$$\mathfrak{A}^T = T^*(\mathcal{H}) = T^*(\mathcal{H})^{\pi\pi} = T^*(\mathcal{H})^{\tau\tau} = T^*(\mathcal{H})^{\rho\rho}.$$

(2) Let $\mathfrak{A} = (\mathbb{C}I, \mathcal{H}, \mathcal{H}^*)$. Then

$$\mathfrak{A}^\pi = T^*(\mathcal{H}) \supsetneqq \mathfrak{A}^\tau = \mathfrak{A}^\rho = \{O\},$$

and

$$\mathfrak{A}^T = \mathfrak{A} = \mathfrak{A}^{\pi\pi} \subsetneqq \mathfrak{A}^{\tau\tau} = \mathfrak{A}^{\rho\rho} = T^*(\mathcal{H}).$$

(3) Let $\mathfrak{A} = (0, \mathcal{H}, \mathcal{H}^*)$. Then we have

$$\begin{aligned}\mathfrak{A}^\pi &= T^*(\mathcal{H}) \supsetneqq \mathfrak{A}^\tau = \mathfrak{A} \supsetneqq \mathfrak{A}^\rho = \{O\}\\ \mathfrak{A}^{\pi\pi} &= (\mathbb{C}I, \mathcal{H}, \mathcal{H}^*), \quad \mathfrak{A}^{\tau\tau} = \mathfrak{A}, \quad \mathfrak{A}^{\rho\rho} = T^*(\mathcal{H}),\end{aligned}$$

so that

$$\mathfrak{A}^T = \mathfrak{A} = \mathfrak{A}^{\tau\tau} \subsetneqq \mathfrak{A}^{\pi\pi} \subsetneqq \mathfrak{A}^{\rho\rho}.$$

(4) Let $\mathfrak{A}_0$ be a nondegenerate C^*-algebra on $\mathcal{H}$, and g a non-zero element of $\mathcal{H}$. We define a CT^*-algebra $\mathfrak{A}$ by $\{(A_0, A_0 g, (A_0 g)^*); \quad A_0 \in \mathfrak{A}_0\}$. Then we see that

$$\begin{aligned}\mathfrak{A}^\pi &\supsetneqq \mathfrak{A}^\tau = \mathfrak{A}^\rho = \{(K_0, K_0 g, (K_0^* g)^*); \quad K_0 \in \mathfrak{A}_0'\},\\ \mathfrak{A}^{\pi\pi} &= (\mathfrak{A}_0'', \mathcal{H}, \mathcal{H}^*) \supsetneqq \mathfrak{A}^{\tau\tau} = \mathfrak{A}^{\rho\rho} = \{(A_0, A_0 g, (A_0^* g)^*); \quad A_0 \in \mathfrak{A}_0''\}.\end{aligned}$$

Thus

$$\mathfrak{A}^T = \mathfrak{A}^{\tau\tau} = \mathfrak{A}^{\rho\rho} = \bar{\mathfrak{A}}[\tau_{\sigma s}^*] \subsetneqq \mathfrak{A}^{\pi\pi}.$$

Putting

$$\nu(X) = (\lambda(X), \lambda(X^\sharp)^*), \quad X \in T^*(\mathcal{H}),$$

$\nu(T^*(\mathcal{H}))$ is a Hilbert space with inner product:

$$(\nu(X)|\nu(Y)) = (\lambda(X)|\lambda(Y)) + (\lambda(Y^\sharp)|\lambda(X^\sharp))$$

for $X, Y \in T^*(\mathcal{H})$, which is identical with the direct sum $\mathcal{H} \oplus \mathcal{H}^*$ of the Hilbert spaces $\mathcal{H}$ and $\mathcal{H}^*$. The subspaces $\nu(\mathfrak{A})$ and $\nu(\mathfrak{A}^{\rho\rho})$ of the Hilbert space $\nu(T^*(\mathcal{H}))$ have the following property:

Proposition 2.6.5 *Let $\mathfrak{A}$ be a CT^*-algebra on $\mathcal{H}$. Then $\nu(\mathfrak{A})$ is dense in $\nu(\mathfrak{A}^{\rho\rho})$, that is,*

$$\nu(\mathfrak{A}^{\rho\rho}) \subset [\nu(\mathfrak{A})],$$

where $[\nu(\mathfrak{A})]$ is the closure of $\nu(\mathfrak{A})$ in the Hilbert space $\nu(T^*(\mathcal{H}))$ under the norm topology.

This will be used in the proof of the von Neumann type density theorem in Chap. 3. In order to prove Proposition 2.6.5 we prepare some lemmas.

Lemma 2.6.6 *Let $[\lambda(\mathfrak{A})]$ be the closure of $\lambda(\mathfrak{A})$ in the Hilbert space $\mathcal{H}$ under the norm topology. Then*

$$\lambda(\mathfrak{A}^{\rho\rho}) \subset [\lambda(\mathfrak{A})].$$

Proof Let $C_{\mathfrak{A}}$ be the projection on $[\lambda(\mathfrak{A})]$. Since $\pi(\mathfrak{A})[\lambda(\mathfrak{A})] \subset [\lambda(\mathfrak{A})]$, we have $C_{\mathfrak{A}} \in \pi(\mathfrak{A})'$, and so $(I - C_{\mathfrak{A}})^\sim := (I - C_{\mathfrak{A}}, 0, 0) \in \mathfrak{A}^\rho$. Hence, since

$$(I - C_{\mathfrak{A}})\lambda(A) = \pi(A)\lambda((I - C_{\mathfrak{A}})^\sim) = 0$$

for all $A \in \mathfrak{A}^{\rho\rho}$, it follows that $C_{\mathfrak{A}}\lambda(A) = \lambda(A)$ for all $A \in \mathfrak{A}^{\rho\rho}$. Therefore, $\lambda(A) \in [\lambda(\mathfrak{A})]$ for all $A \in \mathfrak{A}^{\rho\rho}$. This completes the proof.

Lemma 2.6.7 *Let J be an Hermitian unitary operator on the Hilbert space $\nu(T^*(\mathcal{H}))$ defined by*

$$\begin{pmatrix} I & 0 \\ 0 & -I \end{pmatrix}.$$

Then,

$$\nu(\mathfrak{A}^\rho) \subset J\nu(\mathfrak{A})^\perp \quad \textit{and} \quad \nu(\mathfrak{A}^{\rho\rho}) \subset J\nu(\mathfrak{A}^\rho)^\perp,$$

where $\nu(\mathfrak{A})^\perp$ and $\nu(\mathfrak{A}^\rho)^\perp$ are orthogonal complements of $\nu(\mathfrak{A})$ and $\nu(\mathfrak{A}^\rho)$ in the Hilbert space $\nu(T^*(\mathcal{H}))$, respectively.

Proof Take an arbitrary $K \in \mathfrak{A}^\rho$. Then we have

$$\begin{aligned}(J\nu(K)|\nu(A)) &= \big((\lambda(K), -\lambda(K^\sharp)^*)|(\lambda(A), \lambda(A^\sharp)^*)\big)\\ &= (\lambda(K)|\lambda(A)) - (\lambda(A^\sharp)|\lambda(K^\sharp))\\ &= 0\end{aligned}$$

for all $A \in \mathfrak{A}$. Hence, $\nu(\mathfrak{A}^\rho) \subset J\nu(\mathfrak{A})^\perp$. Applying the same argument to $\mathfrak{A}^\rho$ instead of $\mathfrak{A}$, we can show that $\nu(\mathfrak{A}^{\rho\rho}) \subset J\nu(\mathfrak{A}^\rho)^\perp$.

Lemma 2.6.8 *Let* $\mathcal{H}_0 = [\lambda(\mathfrak{A})] \oplus [\lambda^*(\mathfrak{A})]$. *Then* $\nu(\mathfrak{A}^\rho) \cap \mathcal{H}_0$ *is dense in* $J\nu(\mathfrak{A})^\perp \cap \mathcal{H}_0$.

Proof Take an arbitrary $(x, y^*) \in J\nu(\mathfrak{A})^\perp \cap \mathcal{H}_0$, that is, $x,\ y \in [\lambda(\mathfrak{A})]$ and

$$(\lambda(A)|y) = (x|\lambda(A^\sharp)) \tag{2.6.5}$$

for all $A \in \mathfrak{A}$. We now define operators X_0 and Y_0 on $\lambda(\mathfrak{A})$ by

$$X_0\lambda(A) = \pi(A)x,, \quad Y_0\lambda(A) = \pi(A)y, \quad A \in \mathfrak{A}. \tag{2.6.6}$$

Then it follows from (2.6.5) that

$$\begin{aligned}(\lambda(B)|Y_0\lambda(A)) &= (\lambda(A^\sharp B)|y)\\ &= (x|\lambda(B^\sharp A))\\ &= (X_0\lambda(B)|\lambda(A))\end{aligned}$$

for all $A,\ B \in \mathfrak{A}$. Furthermore, since the closed subspace $[\lambda(\mathfrak{A})]$ of $\mathcal{H}$ is invariant under the action $\pi(\mathfrak{A})$, X_0 and Y_0 are well-defined closable operators in the Hilbert space $[\lambda(\mathfrak{A})]$ satisfying

$$Y_0 \subset X_0^* \quad \text{and} \quad X_0 \subset Y_0^*. \tag{2.6.7}$$

We denote by $\bar{X}_0$ and $\bar{Y}_0$ their closures of X_0 and Y_0 in the Hilbert space $[\lambda(\mathfrak{A})]$, respectively. For closable operators and closed operators in a Hilbert space refer to Appendix B.1. Using the projection $C_{\mathfrak{A}} :=$ Proj $[\lambda(\mathfrak{A})]$ in $\pi(\mathfrak{A})'$, we can define closed operators $\bar{X}_0C_{\mathfrak{A}}$ and $\bar{Y}_0C_{\mathfrak{A}}$ in the Hilbert space $\mathcal{H}$. Then we see that

$$\bar{X}_0C_{\mathfrak{A}} = C_{\mathfrak{A}}\bar{X}_0C_{\mathfrak{A}} \quad \text{and} \quad \bar{Y}_0C_{\mathfrak{A}} = C_{\mathfrak{A}}\bar{Y}_0C_{\mathfrak{A}} \tag{2.6.8}$$

and

$$\bar{X}_0C_{\mathfrak{A}} \text{ and } \bar{Y}_0C_{\mathfrak{A}} \text{ are affiliated with } \pi(\mathfrak{A})', \tag{2.6.9}$$

namely $A_0(\bar{X}_0 C_{\mathfrak{A}}) \subset (\bar{X}_0 C_{\mathfrak{A}})A_0$ and $A_0(\bar{Y}_0 C_{\mathfrak{A}}) \subset (\bar{Y}_0 C_{\mathfrak{A}})A_0$ for all $A_0 \in \pi(\mathfrak{A})''$. Indeed, (2.6.8) is trivial. We will show (2.6.9). It follows immediately from (2.6.6) that $\pi(A)(\bar{X}_0 C_{\mathfrak{A}}) \subset (\bar{X}_0 C_{\mathfrak{A}})\pi(A)$ for all $A \in \mathfrak{A}$ and $B_0(\bar{X}_0 C_{\mathfrak{A}}) \subset (\bar{X}_0 C_{\mathfrak{A}})B_0$ for all $B_0 \in \overline{\pi(\mathfrak{A})}[\tau_s^\pi]$ (the strong closure of $\pi(\mathfrak{A})$), which implies by the von Neumann density theorem (Theorem C.1 in Appendix C) that $A_0(\bar{X}_0 C_{\mathfrak{A}}) \subset (\bar{X}_0 C_{\mathfrak{A}})A_0$ for all $A_0 \in \pi(\mathfrak{A})''$. Similarly, $A_0(\bar{Y}_0 C_{\mathfrak{A}}) \subset (\bar{Y}_0 C_{\mathfrak{A}})A_0$ for all $A_0 \in \pi(\mathfrak{A})''$. Furthermore, it follows from (2.6.7) that

$$\bar{Y}_0 C_{\mathfrak{A}} = C_{\mathfrak{A}} \bar{Y}_0 C_{\mathfrak{A}} \subset C_{\mathfrak{A}} X_0^* C_{\mathfrak{A}} \subset (C_{\mathfrak{A}} \bar{X}_0 C_{\mathfrak{A}})^* = (\bar{X}_0 C_{\mathfrak{A}})^*. \tag{2.6.10}$$

Now, write

$$X = (\bar{X}_0 C_{\mathfrak{A}}, x, y^*)$$

and $\pi(X) = \bar{X}_0 C_{\mathfrak{A}}$, $\lambda(X) = x$ and $\lambda(X^\sharp) = y$. By (2.6.9) and (2.6.10) we have

$$\begin{aligned}
\pi(X)\pi(A) &\supset \pi(A)\pi(X),\\
\pi(X)\lambda(A) &= \bar{X}_0 C_{\mathfrak{A}} \lambda(A) = \pi(A)x = \pi(A)\lambda(X),\\
\pi(A)^*\lambda(X^\sharp) &= \pi(A)^* y = (\bar{Y}_0 C_{\mathfrak{A}})\lambda(A^\sharp)\\
&= (\bar{X}_0 C_{\mathfrak{A}})^*\lambda(A^\sharp) = \pi(X)^*\lambda(A^\sharp)
\end{aligned} \tag{2.6.11}$$

for all $A \in \mathfrak{A}$. This means that X may regard as an *unbounded trio-observable* in $\mathcal{H}$ which commutes with the T^*-algebra $\mathfrak{A}$. Using the polar decomposition of $\pi(X)$ and the spectral resolutions of $\pi(X)\pi(X)^*$ and $\pi(X)^*\pi(X)$, we can find a sequence $\{X_n\}$ in $\mathfrak{A}^\rho$ satisfying $\lim_{n\to\infty} \nu(X_n) = (x, y^*)$. Let $\pi(X) = U|\pi(X)|$ be the polar decomposition of $\pi(X)$. Then, $\pi(X)^* = U^*|\pi(X)^*|$ is the polar decomposition of $\pi(X)^*$. Let $|\pi(X)|^2 = \pi(X)^*\pi(X) = \int_0^\infty t dE(t)$ and $|\pi(X)^*|^2 = \pi(X)\pi(X)^* = \int_0^\infty t dF(t)$ be the spectral resolutions of $|\pi(X)|^2$ and $|\pi(X)^*|^2$, respectively. For the polar decomposition of a densely defined closed operator and the spectral resolution of a self-adjoint operator, see Theorems B.10 and B.5 in Appendix B, respectively. Then we have

$$|\pi(X)| = \int_0^\infty \sqrt{t} dE(t) \text{ and } |\pi(X)^*| = \int_0^\infty \sqrt{t} dF(t), \tag{2.6.12}$$

and by (2.6.9) that

$$U,\ E(t),\ F(t) \in \pi(\mathfrak{A})' \text{ and } F(t) = UE(t)U^* \tag{2.6.13}$$

for all $t \in [0, \infty)$. Since $\overline{F(n)\pi(X)} = |\pi(X)^*|F(n)U \in B(\mathcal{H})$ for all $n \in \mathbb{N}$, we can define an element $X_n \in T^*(\mathcal{H})$ by

$$X_n = (\overline{F(n)\pi(X)}, F(n)x, (E(n)y)^*), \quad n \in \mathbb{N}.$$

Since $C_{\mathfrak{A}}\pi(X) \subset \pi(X)C_{\mathfrak{A}}$ and $C_{\mathfrak{A}}\pi(X)^* \subset \pi(X)^*C_{\mathfrak{A}}$ by (2.6.8), we get that $E(n)[\lambda(\mathfrak{A})] \subset [\lambda(\mathfrak{A})]$ and $F(n)[\lambda(\mathfrak{A})] \subset [\lambda(\mathfrak{A})]$ for all $n \in \mathbb{N}$. Hence it follows that

$$\nu(X_n) = (F(n)x, (E(n)y)^*) \in \mathcal{H}_0, \tag{2.6.14}$$

and from (2.6.11) and (2.6.12) that

$$\begin{aligned}
\pi(X_n) &= |\pi(X)^*|F(n)U \in \pi(\mathfrak{A})', \\
\pi(X_n)\lambda(A) &= F(n)\pi(X)\lambda(A) \\
&= F(n)\pi(A)x \\
&= \pi(A)F(n)x,
\end{aligned}$$

$$\begin{aligned}
\pi(X_n)^*\lambda(A^\sharp) &= U^*|\pi(X)^*|F(n)\lambda(A^\sharp) \\
&= U^*|\pi(X)^*|UE(n)U^*\lambda(A^\sharp) \\
&= |\pi(X)|E(n)U^*\lambda(A^\sharp) \\
&= E(n)\pi(X)^*\lambda(A^\sharp) \\
&= E(n)\pi(A)^*y \\
&= \pi(A)^*E(n)y
\end{aligned}$$

for all $A \in \mathfrak{A}$. Therefore, $X_n \in \mathfrak{A}^\tau$. Furthermore, we can prove that

$$(\lambda(X_n)|\lambda(A^\sharp)) = (\lambda(A)|\lambda(X_n^\sharp)) \tag{2.6.15}$$

for all $A \in \mathfrak{A}$. Indeed, for $\alpha < 0$ take a continuous bounded function f on $[0, \infty)$:

$$f(t) = \frac{1}{t-\alpha} = -\frac{1}{\alpha}\left(1 - \frac{t}{t-\alpha}\right), \quad t \in [0, \infty).$$

Since

$$\begin{aligned}
f(|\pi(X)^*|^2) &= \int_0^\infty \frac{1}{t-\alpha} dF(t), \\
|\pi(X)^*|^2 f(|\pi(X)^*|^2) &= \int_0^\infty \frac{t}{t-\alpha} dF(t), \\
\pi(X)^* f(|\pi(X)^*|^2) &= U^*\left(\int_0^\infty \frac{\sqrt{t}}{t-\alpha} dF(t)\right) \quad \text{by (2.6.2)},
\end{aligned}$$

(see Theorem B.6 in Appendix B), they are bounded, and similarly $f(|\pi(X)|^2)$ and $|\pi(X)|^2 f(|\pi(X)|^2)$ are both bounded. Hence it follows from (2.6.11) and (2.6.13) that

$$\begin{aligned}
(f(|\pi(X)^*|^2)x|\lambda(A^\sharp)) &= -\frac{1}{\alpha}(x|\lambda(A^\sharp)) + \frac{1}{\alpha}(|\pi(X)^*|^2 f(|\pi(X)^*|^2)x|\lambda(A^\sharp)) \\
&= -\frac{1}{\alpha}(\lambda(A)|y) + \frac{1}{\alpha}(\pi(X)^* f(|\pi(X)^*|^2)x|\pi(X)^*\lambda(A^\sharp)) \\
&= -\frac{1}{\alpha}(\lambda(A)|y) + \frac{1}{\alpha}(\pi(X)^* f(|\pi(X)^*|^2)x|\pi(A)^* y) \\
&= -\frac{1}{\alpha}(\lambda(A)|y) + \frac{1}{\alpha}(\pi(X)^* f(|\pi(X)^*|^2)\pi(A)x|y) \\
&= -\frac{1}{\alpha}(\lambda(A)|y) + \frac{1}{\alpha}(\pi(X)^* f(|\pi(X)^*|^2)\pi(X)\lambda(A)|y) \\
&= -\frac{1}{\alpha}(\lambda(A)|y) \\
&\quad + \frac{1}{\alpha}\left(|\pi(X)|U^*\left(\int_0^\infty t dF(t)\right)U|\pi(X)|\lambda(A)|y\right) \\
&= -\frac{1}{\alpha}(\lambda(A)|y) \\
&\quad + \frac{1}{\alpha}\left(|\pi(X)|\left(\int_0^\infty t dU^*F(t)U\right)|\pi(X)|\lambda(A)|y\right) \\
&= -\frac{1}{\alpha}(\lambda(A)|y) \\
&\quad + \frac{1}{\alpha}\left(|\pi(X)|\left(\int_0^\infty t dE(t)\right)|\pi(X)|\lambda(A)|y\right) \\
&= -\frac{1}{\alpha}(\lambda(A)|y) + \frac{1}{\alpha}(|\pi(X)| f(|\pi(X)|^2)|\pi(X)|\lambda(A)|y) \\
&= (\lambda(A)| -\frac{1}{\alpha}(I - |\pi(X)|^2) f(|\pi(X)|^2)y) \\
&= (\lambda(A)|f(|\pi(X)|^2)y) \qquad (2.6.16)
\end{aligned}$$

for all $A \in \mathfrak{A}$. Let χ_n be an indicator function of the closed interval $[0, n]$ on $[0, \infty)$. Since

$$\chi_n(t) = \frac{1}{\log(-(n-\alpha)|\alpha)} \int_0^n f(t)dt = 1, \quad t \in [0, n],$$

it follows from (2.6.16) that

$$\begin{aligned}(\lambda(X_n)|\lambda(A^\sharp)) &= (F(n)x|\lambda(A^\sharp)) \\ &= (\chi_n(|\pi(X)^*|^2)x|\lambda(A^\sharp)) \\ &= \frac{1}{\log(-(n-\alpha)|\alpha)}\int_0^n (f(|\pi(X)^*|^2)x|\lambda(A^\sharp))dt \\ &= \frac{1}{\log(-(n-\alpha)|\alpha)}\int_0^n (\lambda(A)|f(|\pi(X)|^2)y)dt \\ &= (\lambda(A)|\chi_n(|\pi(X)|^2)y) \\ &= (\lambda(A)|E(n)y) \\ &= (\lambda(A)|\lambda(X_n^\sharp))\end{aligned}$$

for all $A \in \mathfrak{A}$, which proves (2.6.15). Thus we get that $X_n \in \mathfrak{A}^\rho$ and $\nu(X_n) \in \mathcal{H}_0$ by (2.6.14), and

$$\begin{aligned}\lim_{n\to\infty} \nu(X_n) &= \lim_{n\to\infty}(F(n)x, (E(n)y)^*) \\ &= (x, y^*) \in J\nu(\mathfrak{A})^\perp \cap \mathcal{H}_0.\end{aligned}$$

Thus, $\nu(\mathfrak{A}^\rho) \cap \mathcal{H}_0$ is dense in $J\nu(\mathfrak{A})^\perp \cap \mathcal{H}_0$. This completes the proof.

The Proof of Proposition 2.6.5 By Lemma 2.6.6 we have

$$\nu(\mathfrak{A}^{\rho\rho}) \subset \mathcal{H}_0. \tag{2.6.17}$$

Take an arbitrary $\xi \in \nu(\mathfrak{A}^{\rho\rho})$. By Lemma 2.6.7 we can prove that

$$\begin{aligned}\nu(\mathfrak{A}^{\rho\rho}) &\subset J\nu(\mathfrak{A}^\rho)^\perp \\ &\subset J(J\nu(\mathfrak{A})^\perp \cap \mathcal{H}_0)^\perp \\ &= (\nu(\mathfrak{A})^\perp \cap \mathcal{H}_0)^\perp,\end{aligned}$$

and by (2.6.17) that

$$\xi \in \mathcal{H}_0 \quad \text{and} \quad \xi \in (\nu(\mathfrak{A})^\perp \cap \mathcal{H}_0)^\perp. \tag{2.6.18}$$

Take an arbitrary $\eta \in \nu(\mathfrak{A})^\perp$. Then η is decomposed into

$$\eta = \eta_1 + \eta_2, \quad \eta_1 \in \mathcal{H}_0, \ \eta_2 \in \mathcal{H}_0^\perp.$$

By (2.6.17) we have

$$\eta_2 \in \mathcal{H}_0^{\perp} \subset \nu(\mathfrak{A}^{\rho\rho})^{\perp} \subset \nu(\mathfrak{A})^{\perp};$$

hence

$$\eta_1 = \eta - \eta_2 \in \nu(\mathfrak{A})^{\perp} \cap \mathcal{H}_0.$$

It therefore follows from (2.6.18) that

$$\begin{aligned}(\xi|\eta) &= (\xi|\eta_1) + (\xi|\eta_2)\\ &= 0,\end{aligned}$$

which implies that $\xi \in \nu(\mathfrak{A})^{\perp\perp} = [\nu(\mathfrak{A})]$. Thus $\nu(\mathfrak{A})$ is dense in $\nu(\mathfrak{A}^{\rho\rho})$. This completes the proof.

Notes As for the basic definitions and the fundamental theorems (spectral resolution, functional calculus for a self-adjoint operator, polar decomposition of a densely defined closed operator in a Hilbert space), refer the reader to Appendix B.

Chapter 3
Density Theorems

In this chapter we will generalize the von Neumann density theorem and the Kaplansky density theorem (see Appendix C) which are the fundamental results of the theory of von Neumann algebras to the case of observable algebras. They play an important role in studies of observable algebras.

3.1 Von Neumann Type Density Theorem

The purpose of this section is to prove the following von Neumann type density theorem.

Theorem 3.1.1 (von Neumann Type Density Theorem) *Let $\mathfrak{A}$ be a T^*-algebra on $\mathcal{H}$. Then,*

$$\bar{\mathfrak{A}}[\tau^*_{\sigma s}] = \mathfrak{A}^T := \mathfrak{A}^{\pi\pi} \cap \mathfrak{A}^{\tau\tau} \cap \mathfrak{A}^{\rho\rho}.$$

Since $\bar{\mathfrak{A}}[\tau^*_{\sigma s}]$ coincides with the closure of $\bar{\mathfrak{A}}[\tau_u]$ under the σ-strong* topology, we may assume without loss of generality that $\mathfrak{A}$ is a CT^*-algebra. We shall prove this theorem after some preparations.

Let $\mathcal{H}$ and $\mathcal{K}$ be Hilbert spaces. For $A \in T^*(\mathcal{H})$ and $B \in T^*(\mathcal{K})$ we define the tensor product $A \otimes B$ of A and B by

$$A \otimes B = \left(\pi(A) \otimes \pi(B), \lambda(A) \otimes \lambda(B), (\lambda(A^\sharp) \otimes \lambda(B^\sharp))^*\right).$$

Let $l^2 = l^2(\mathbb{Z}_+)$, $\mathbb{Z}_+ = \{0, 1, 2, \ldots\}$ and $e_k = (0, \ldots, 0, \overset{(k)}{1}, 0, \ldots)$, $k \in \mathbb{Z}_+$. Put

$$G = (I, e_0, e_0^*) \in T^*(l^2).$$

A. Inoue, *Tomita's Lectures on Observable Algebras in Hilbert Space*,
Lecture Notes in Mathematics 2285, https://doi.org/10.1007/978-3-030-68893-6_3

Then, we have, for $A \in T^*(\mathcal{H})$,

$$A \otimes G = (\pi(A) \otimes I, \lambda(A) \otimes e_0, (\lambda(A^\sharp) \otimes e_0)^*),$$

and the map:

$$\psi : \quad A \mapsto A \otimes G$$

is a σ-strongly* homeomorphic $*$-isomorphism of $T^*(\mathcal{H})$ into $T^*(\mathcal{H} \otimes l^2)$. We now denote by $\mathcal{H}_{\mathbb{Z}_+}$ and $\mathcal{H}_{\mathbb{N}}$ the direct sums $\oplus_{k=0}^{\infty}\mathcal{H}_k$ and $\oplus_{k=1}^{\infty}\mathcal{H}_k$ of $\{\mathcal{H}_k\}_{k\in\mathbb{Z}_+}$ and $\{\mathcal{H}_k\}_{k\in\mathbb{N}}$, where $\mathcal{H}_k = \mathcal{H} \otimes e_k$, $k \in \mathbb{Z}_+$, respectively. Then we can regard an element $g = (g_k)$ of the Hilbert space $\mathcal{H}_{\mathbb{Z}_+}$ (resp. $\mathcal{H}_{\mathbb{N}}$) as an element $\sum_{k=0}^{\infty} g_k \otimes e_k$ (resp. $\sum_{k=0}^{\infty} g_k \otimes e_k$, $e_0 = 0$) of the Hilbert space $\mathcal{H} \otimes l^2$. Then, for $g = (g_k) \in \mathcal{H}_{\mathbb{N}}$ and $A \in T^*(\mathcal{H})$ we obtain

$$T_g(A \otimes G) = \Big(\pi(A) \otimes I, \lambda(A) \otimes e_0 + \sum_{k=1}^{\infty} \pi(A)g_k \otimes e_k, (\lambda(A^\sharp) \otimes e_0 + \sum_{k=1}^{\infty} \pi(A)^* g_k \otimes e_k)^*\Big),$$

so that

$$\|\lambda(T_g(A \otimes G))\| = \left(\sum_{k=1}^{\infty} \|\pi(A)g_k\|^2 + \|\lambda(A)\|^2\right)^{\frac{1}{2}}$$

and

$$\|\nu(T_g(A \otimes G))\| = \left(\sum_{k=1}^{\infty} \|\pi(A)g_k\|^2 + \sum_{k=1}^{\infty} \|\pi(A)^* g_k\|^2 + \|\lambda(A)\|^2 + \|\lambda(A^\sharp)\|^2\right)^{\frac{1}{2}}$$

for all $A \in T^*(\mathcal{H})$, which implies the following

Lemma 3.1.2 *The σ-strong* topology $\tau^*_{\sigma s}$ on $T^*(\mathcal{H})$ is the weakest topology for which the map:*

$$A \in T^*(\mathcal{H}) \mapsto \nu(T_g(A \otimes G)) \in (\mathcal{H} \otimes l^2) \oplus (\mathcal{H} \otimes l^2)^*$$

is continuous for every $g \in \mathcal{H}_{\mathbb{N}}$.

For every element X of $T^*(\mathcal{H} \otimes l^2)$, $\pi(X)$ and $\lambda(X)$ are represented as

$$\begin{aligned}\pi(X) &= (\pi_{jk}(X)) \\ &= \sum_{j,k\in\mathbb{Z}_+} \pi_{jk}(X) \otimes e_{jk} \in B(\mathcal{H} \otimes l^2)\end{aligned}$$

and

$$\lambda(X) = \sum_{k \in \mathbb{Z}_+} \lambda_k(X) \otimes e_k \in \mathcal{H} \otimes l^2,$$

where $\pi_{jk}(X) \in B(\mathcal{H})$ and $\lambda_k(X) \in \mathcal{H}$ for $j,\ k \in \mathbb{Z}_+$.

For the commutants $(\mathfrak{A} \otimes G)^{\pi}$, $(\mathfrak{A} \otimes G)^{\tau}$ and $(\mathfrak{A} \otimes G)^{\rho}$ we have the following

Lemma 3.1.3 *Let $\mathfrak{A}$ be a CT^*-algebra on $\mathcal{H}$. Then the following (1), (2) and (3) hold:*

(1) $(\mathfrak{A} \otimes G)^{\pi} = (\pi(\mathfrak{A})' \otimes B(l^2), \mathcal{H} \otimes l^2, (\mathcal{H} \otimes l^2)^*)$.
(2) $K \in (\mathfrak{A} \otimes G)^{\tau}$ *if and only if*

$$\begin{aligned}\pi(K) &= (\pi_{jk}(K)) \\ &= \sum_{j,k \in \mathbb{Z}_+} \pi_{jk}(K) \otimes e_{jk}\end{aligned}$$

and

$$\lambda(K) = \sum_{k=0}^{\infty} \lambda_k(K) \otimes e_k, \quad \lambda_k(K) \in \mathcal{H}$$

satisfy the following (i)–(iii):

(i) $\pi_{jk}(K) \in \pi(\mathfrak{A})'$ *for* $k,\ j \in \mathbb{Z}_+$.
(ii) $K_{00} := (\pi_{00}(K), \lambda_0(K), \lambda_0(K^{\natural})^*) \in \mathfrak{A}^{\tau}$.
(iii) Take an arbitrary $j \in \mathbb{N}$. *We put*

$$\begin{aligned}K_{j0} &= (\pi_{j0}(K), \lambda_j(K), \lambda_j(K^{\natural})^*), \\ K_{0j} &= (\pi_{0j}(K), \lambda_j(K), \lambda_j(K^{\natural})^*)\end{aligned}$$

and

$$\begin{aligned}|K_{j0}| &= (K_{j0}^{\natural} K_{j0})^{1/2} = \left(|\pi_{j0}(K)|, U_j^* \lambda_j(K), (U_j^* \lambda_j(K))^*\right), \\ |K_{0j}^{\natural}| &= (K_{0j} K_{0j}^{\natural})^{1/2} = \left(|\pi_{0j}(K)^*|, V_j^* \lambda_j(K^{\natural}), (V_j^* \lambda_j(K^{\natural}))^*\right),\end{aligned}$$

where $\pi_{j0}(K) = U_j |\pi_{j0}(K)|$ *and* $\pi_{0j}(K)^* = V_j |\pi_{0j}(K)^*|$ *are the polar decompositions of* $\pi_{j0}(K)$ *and* $\pi_{0j}(K)^*$, *respectively. Then,*

$$|K_{j0}|,\ |K_{j0}^{\natural}| \in \mathfrak{A}^{\tau}$$

and

$$\lambda_j(K) = U_j\lambda(|K_{j0}|) + x_j,$$
$$\lambda_j(K^\sharp) = V_j\lambda(|K^\sharp_{0j}|) + y_j,$$

where x_j and y_j are elements of $\mathcal{H}$ satisfying $\pi(A)x_j = 0$ and $\pi(A)y_j = 0$ for all $A \in \mathfrak{A}$.

(3) $(\mathfrak{A} \otimes G)^\rho = \{K \in (\mathfrak{A} \otimes G)^\tau; \quad K_{00} \in \mathfrak{A}^\rho\}$.

Proof

(1) This is trivial.
(2) Take an arbitrary $K \in (\mathfrak{A} \otimes G)^\tau$. Then we have

$$\pi(K)(\pi(A) \otimes I) = (\pi(A) \otimes I)\pi(K), \tag{3.1.1}$$

$$\pi(K)(\lambda(A) \otimes e_0) = (\pi(A) \otimes I)\lambda(K), \tag{3.1.2}$$

$$\pi(K)^*(\lambda(A^\sharp) \otimes e_0) = (\pi(A)^* \otimes I)\lambda(K^\sharp) \tag{3.1.3}$$

for all $A \in \mathfrak{A}$. By (3.1.1) we get (i): $\pi_{jk}(K) \in \pi(\mathfrak{A})'$ for all $j,\ k \in \mathbb{Z}_+$, and so (i) holds. By (3.1.2) and (3.1.3) we can prove that

$$\sum_{j=0}^{\infty} \pi_{j0}(K)\lambda(A) \otimes e_j = \sum_{j=0}^{\infty} \pi(A)\lambda_j(K) \otimes e_j,$$

$$\sum_{j=0}^{\infty} \pi_{0j}(K)^*\lambda(A) \otimes e_j = \sum_{j=0}^{\infty} \pi(A)\lambda_j(K^\sharp) \otimes e_j,$$

so that

$$\pi_{j0}(K)\lambda(A) = \pi(A)\lambda_j(K), \tag{3.1.4}$$

$$\pi_{0j}(K)^*\lambda(A) = \pi(A)\lambda_j(K^\sharp) \tag{3.1.5}$$

for all $j \in \mathbb{Z}_+$. In particular, we have

$$\pi_{00}(K) \in \pi(\mathfrak{A})',$$
$$\pi_{00}(K)\lambda(A) = \pi(A)\lambda_0(K),$$
$$\pi_{00}(K)^*\lambda(A^\sharp) = \pi(A)^*\lambda_0(K^\sharp),$$

which implies (ii):

$$K_{00} = (\pi_{00}(K), \lambda_0(K), \lambda_0(K^\sharp)^*) \in \mathfrak{A}^\tau.$$

However, we cannot get $K_{j0} \in \mathfrak{A}^\tau$ from (3.1.4) and (3.1.5) because the equality: $\pi_{j0}(K)^*\lambda(A^\sharp) = \pi(A)^*\lambda_j(K^\sharp)$ does not hold. Similarly, we cannot get $K^\sharp_{0j} \in \mathfrak{A}^\tau$. From this, we consider the absolutes $|K_{j0}|$ and $|K^\sharp_{0j}|$ of K_{j0} and $K^\sharp_{0j}$, and show that they are elements of $\mathfrak{A}^\tau$. Since U_j, $|\pi_{j0}(K)| \in \pi(\mathfrak{A})'$, it follows from (3.1.4) that

$$\begin{aligned}|\pi_{j0}(K)|\lambda(A) &= U_j^*\pi_{j0}(K)\lambda(A)\\ &= U_j^*\pi(A)\lambda_j(K)\\ &= \pi(A)U_j^*\lambda_j(K)\end{aligned}$$

for all $A \in \mathfrak{A}$. Therefore

$$|K_{j0}| = \Big(|\pi_{j0}(K)|, U_j^*\lambda_j(K), (U_j^*\lambda_j(K))^*\Big) \in \mathfrak{A}^\tau,$$

and $\lambda_j(K) = U_j\lambda(|K_{j0}|) + x_j$, where $x_j := (I - U_jU_j^*)\lambda_j(K)$. Then, by (3.1.4) we obtain

$$\begin{aligned}\pi(A)x_j &= (I - U_jU_j^*)\pi(A)\lambda_j(K)\\ &= (I - U_jU_j^*)\pi_{j0}(K)\lambda(A)\\ &= 0\end{aligned}$$

for all $A \in \mathfrak{A}$. Furthermore, we can prove in the same way as K_{j0} that $|K^\sharp_{0j}| = \Big(|\pi_{0j}(K)^*|, V_j^*\lambda_j(K^\sharp), (V_j^*\lambda_j(K^\sharp))^*\Big) \in \mathfrak{A}^\tau$ and $\lambda_j(K^\sharp) = V_j\lambda(|K^\sharp_{0j}|) + y_j$, where $y_j := (I - V_jV_j^*)\lambda_j(K^\sharp)$. Then $\pi(A)y_j = 0$ for all $A \in \mathfrak{A}$. Thus (iii) holds. Suppose, conversely, that $K \in T^*(\mathcal{H} \otimes l^2)$ satisfies statements (i)–(iii). Then, since $\pi(A)x_j = 0$ and $\pi(A)y_j = 0$ for all $A \in \mathfrak{A}$ and $j \in \mathbb{N}$, we have

$$\begin{aligned}(\pi(A) \otimes I)\lambda(K) &= (\pi(A) \otimes I)(\sum_{j=0}^{\infty}\lambda_j(K) \otimes e_j)\\ &= \pi(A)\lambda_0(K) \otimes e_0 + \sum_{j=1}^{\infty}\pi(A)\lambda_j(K) \otimes e_j\\ &= \pi_{00}(K)\lambda(A) \otimes e_0 + \sum_{j=1}^{\infty}\pi(A)(U_j\lambda(|K_{j0}|) + x_j) \otimes e_j\end{aligned}$$

$$
\begin{aligned}
&= \pi_{00}(K)\lambda(A) \otimes e_0 + \sum_{j=1}^{\infty} \pi(A)U_j\lambda(|K_{j0}|) \otimes e_j \\
&= \pi_{00}(K)\lambda(A) \otimes e_0 + \sum_{j=1}^{\infty} U_j\pi(|K_{j0}|)\lambda(A) \otimes e_j \\
&= \pi_{00}(K)\lambda(A) \otimes e_0 + \sum_{j=1}^{\infty} \pi_{j0}(K)\lambda(A) \otimes e_j \\
&= \sum_{j=0}^{\infty} \pi_{j0}(K)\lambda(A) \otimes e_j \\
&= \pi(K)(\lambda(A) \otimes e_0)
\end{aligned}
$$

and

$$
\begin{aligned}
(\pi(A)^* \otimes I)\lambda(K^\sharp) &= \sum_{j=0}^{\infty} \pi(A)^*\lambda_j(K^\sharp) \otimes e_j \\
&= \pi(A)^*\lambda_0(K^\sharp) + \sum_{j=1}^{\infty} \pi(A)^*(V_j\lambda(|K_{0j}^\sharp|) + y_j) \otimes e_j \\
&= \pi_{00}(K)^*\lambda(A^\sharp) \otimes e_0 + \sum_{j=1}^{\infty} V_j\pi(|K_{0j}^\sharp|)\lambda(A^\sharp) \otimes e_j \\
&= \sum_{j=0}^{\infty} \pi_{0j}(K)^*\lambda(A^\sharp) \otimes e_j \\
&= \pi(K)^*(\lambda(A^\sharp) \otimes e_0).
\end{aligned}
$$

Therefore, $K \in (\mathfrak{A} \otimes G)^\tau$.

(3) This follows from the following:

$$
\begin{aligned}
K \in (\mathfrak{A} \otimes G)^\rho \quad &\text{iff } K \in (\mathfrak{A} \otimes G)^\tau \\
&\quad \text{and } (\lambda(A \otimes G)|\lambda(K^\sharp)) = (\lambda(K)|\lambda((A \otimes G)^\sharp)) \text{ for all } A \in \mathfrak{A} \\
&\text{iff } K \in (\mathfrak{A} \otimes G)^\tau \\
&\quad \text{and } (\lambda(A)|\lambda(K_{00}^\sharp)) = (\lambda(K_{00})|\lambda(A^\sharp)) \text{ for all } A \in \mathfrak{A} \\
&\text{iff } K \in (\mathfrak{A} \otimes G)^\tau \text{ and } K_{00} \in \mathfrak{A}^\rho.
\end{aligned}
$$

This completes the proof.

Lemma 3.1.4 *Let $\mathfrak{A}$ be a CT^*-algebra on $\mathcal{H}$ and $\mathfrak{A}^T = \mathfrak{A}^{\pi\pi} \cap \mathfrak{A}^{\tau\tau} \cap \mathfrak{A}^{\rho\rho}$. Then*

$$\mathfrak{A}^T \otimes G = (\mathfrak{A} \otimes G)^T.$$

Proof By the equality:

$$(\mathfrak{A} \otimes G)^{\pi\pi} = \left(\pi(\mathfrak{A})'' \otimes I, \mathcal{H} \otimes l^2, (\mathcal{H} \otimes l^2)^*\right), \tag{3.1.6}$$

we have

$$\mathfrak{A}^T \otimes G \subset (\mathfrak{A} \otimes G)^{\pi\pi}. \tag{3.1.7}$$

Suppose $A \in \mathfrak{A}^T$ and $K \in (\mathfrak{A} \otimes G)^\tau$. Then, since $A \in \mathfrak{A}^{\tau\tau}$ and $\pi(\mathfrak{A})x_j = 0$ for all $j \in \mathbb{Z}_+$, where x_j is as in Lemma 3.1.3, and $P_{\mathfrak{A}} = P_{\mathfrak{A}^{\tau\tau}}$ by Theorem 4.1.7 in Sect. 4.1, it follows that $P_{\mathfrak{A}^{\tau\tau}} x_j = P_{\mathfrak{A}} x_j = 0$, which implies that $\pi(A)x_j = 0$ for all $j \in \mathbb{Z}_+$. Hence we can show in the same as the proof of Lemma 3.1.3, (2) that

$$(\pi(A) \otimes I)\lambda(K) = \pi(K)\lambda(A \otimes G)$$

and

$$(\pi(A) \otimes I)^*\lambda(K^\sharp) = \pi(K)^*\lambda((A \otimes G)^\sharp).$$

Therefore $A \otimes G \in (\mathfrak{A} \otimes G)^{\tau\tau}$. Thus

$$\mathfrak{A}^T \otimes G \subset (\mathfrak{A} \otimes G)^{\tau\tau}. \tag{3.1.8}$$

Furthermore, for any $A \in \mathfrak{A}^T$ and $K \in (\mathfrak{A} \otimes G)^\rho$, it follow from Lemma 3.1.3 that

$$\begin{aligned}
(\lambda(A \otimes G)|\lambda(K^\sharp)) &= (\lambda(A) \otimes e_0|\lambda(K^\sharp)) \\
&= (\lambda(A)|\lambda(K_0^\sharp)) \\
&= (\lambda(K_0)|\lambda(A^\sharp)) \\
&= (\lambda(K)|\lambda((A \otimes G)^\sharp)).
\end{aligned}$$

Hence, $A \otimes G \in (\mathfrak{A} \otimes G)^{\rho\rho}$, so

$$\mathfrak{A}^T \otimes G \subset (\mathfrak{A} \otimes G)^{\rho\rho}. \tag{3.1.9}$$

By (3.1.7)–(3.1.9) we have

$$\mathfrak{A}^T \otimes G \subset (\mathfrak{A} \otimes G)^T.$$

Conversely, take an arbitrary $X \in (\mathfrak{A} \otimes G)^T$. By (3.1.6) there exists an element X_0 of $\pi(\mathfrak{A})''$ such that $\pi(X) = X_0 \otimes I$. Furthermore, since $XK = KX$ for all $K \in (\mathfrak{A} \otimes G)^\tau$, that is,

$$(X_0 \otimes I)\lambda(K) = \pi(K)\lambda(X),$$
$$(X_0^* \otimes I)\lambda(K^\sharp) = \pi(K)^*\lambda(X^\sharp),$$

we have

$$X_0\lambda_j(K) = \sum_{k=0}^{\infty} \pi_{jk}(K)\lambda_k(X), \tag{3.1.10}$$

$$X_0\lambda_j(K^\sharp) = \sum_{k=0}^{\infty} \pi_{jk}(K)^*\lambda_k(X^\sharp). \tag{3.1.11}$$

For any $K_{00} = (\pi(K_{00}), \lambda(K_{00}), \lambda(K_{00}^\sharp)) \in \mathfrak{A}^\tau$ and $j \in \mathbb{N}$, we define an element $K = (\pi(K), \lambda(K), \lambda(K^\sharp)^*)$ of $T^*(\mathcal{H} \otimes l^2)$ as follows:

$$\pi(K) = \pi(K_{00}) \otimes e_{00} + I \otimes e_{1j}, \quad (j > 0),$$
$$\lambda(K) = \lambda(K_{00}) \otimes e_0,$$
$$\lambda(K^\sharp) = \lambda(K_{00}^\sharp) \otimes e_0.$$

Then, by Lemma 3.1.3, (2) we get $K \in (\mathfrak{A} \otimes G)^\tau$. Take an arbitrary $j > 1$. Since

$$\begin{aligned} X_0\lambda(K_{00}) \otimes e_0 &= (X_0 \otimes I)\lambda(K) \\ &= \pi(K)\lambda(X) \\ &= (\pi(K_{00}) \otimes e_{00} + I \otimes e_{1j})(\sum_k \lambda_k(X) \otimes e_k) \\ &= \pi(K_{00})\lambda_0(X) \otimes e_0 + \lambda_j(X) \otimes e_1, \end{aligned}$$

it follows that

$$X_0\lambda(K_{00}) = \pi(K_{00})\lambda_0(X) \ \text{ and } \ \lambda_j(X) = 0. \tag{3.1.12}$$

Similarly, we see that

$$X_0^*\lambda(K_{00}^\sharp) = \pi(K_{00})^*\lambda_0(X^\sharp) \ \text{ and } \ \lambda_j(X^\sharp) = 0. \tag{3.1.13}$$

Therefore

$$(X_0, \lambda_0(X), \lambda_0(X^\sharp)^*) \in \mathfrak{A}^{\tau\tau}. \tag{3.1.14}$$

Furthermore, it follows from (3.1.12) and (3.1.13) that

$$\begin{aligned}\pi(X) &= X_0 \otimes I,\\ \lambda(X) &= \sum_{j=0}^{\infty} \lambda_j(X) \otimes e_j = \lambda_0(X) \otimes e_0,\\ \lambda(X^\sharp) &= \sum_{j=0}^{\infty} \lambda_j(X^\sharp) \otimes e_j = \lambda_0(X^\sharp) \otimes e_0,\end{aligned}$$

which implies by (3.1.14) that $X \in \mathfrak{A}^{\tau\tau} \otimes G$. Furthermore, we get

$$(\lambda(X)|\lambda(K^\sharp)) = (\lambda(K)|\lambda(X^\sharp))$$

for all $K \in (\mathfrak{A} \otimes G)^\rho$, which implies since $\mathfrak{A}^\rho \otimes G \subset (\mathfrak{A} \otimes G)^\rho$ that

$$(\lambda_0(X)|\lambda(K_0^\sharp)) = (\lambda(K_0)|\lambda_0(X^\sharp))$$

for all $K_0 \in \mathfrak{A}^\rho$. Hence, $(X_0, \lambda_0(X), \lambda_0(X^\sharp)^*) \in \mathfrak{A}^{\rho\rho}$, so that $X \in \mathfrak{A}^{\rho\rho} \otimes G$ and $X \in \mathfrak{A}^T \otimes G$. Thus we have

$$(\mathfrak{A} \otimes G)^T \subset \mathfrak{A}^T \otimes G.$$

This completes the proof.

The Proof of Theorem 3.1.1 By Proposition 2.6.3 and Lemma 3.1.4 we have

$$\begin{aligned}T_g(\mathfrak{A} \otimes G) &\subset T_g(\mathfrak{A}^T \otimes G)\\ &= (T_g(\mathfrak{A} \otimes G))^T\\ &\subset T_g(\mathfrak{A} \otimes G)^{\rho\rho}\end{aligned} \tag{3.1.15}$$

for any $g \in G$. Since $\nu(T_g(\mathfrak{A} \otimes G))$ is dense in $\nu((T_g(\mathfrak{A} \otimes G))^{\rho\rho})$ by Proposition 2.6.5, it follows from (3.1.15) that $\nu(T_g(\mathfrak{A} \otimes G))$ is dense in $\nu(T_g(\mathfrak{A}^T \otimes G))$, which implies by Lemma 3.1.2 that $\mathfrak{A}$ is strongly* dense in $\mathfrak{A}^T$. Furthermore, since $\mathfrak{A}^T$ is strongly* closed in $T^*(\mathcal{H})$, we have $\bar{\mathfrak{A}}[\tau_{\sigma s}^*] = \mathfrak{A}^T$. This completes the proof.

Theorem 3.1.1 proves the following

Corollary 3.1.5 *Let $\mathfrak{A}$ be a T^*-algebra on $\mathcal{H}$. Then $\mathfrak{A}^T$ coincides with each of the closures of $\mathfrak{A}$ under the topologies τ_w, $\tau_{\sigma w}$, τ_s, $\tau_{\sigma s}$, τ_s^* and $\tau_{\sigma s}^*$.*

Proof By Proposition 2.6.2, $\mathfrak{A}^T$ is weakly closed, and furthermore by (2.5.1) we have

$$\bar{\mathfrak{A}}[\tau_{\sigma s}^*] \subset \bar{\mathfrak{A}}[\tau] \subset \bar{\mathfrak{A}}[\tau_w] \subset \mathfrak{A}^T,$$

where τ is one of the topologies $\tau_{\sigma w}$, τ_s, $\tau_{\sigma s}$ and τ_s^*, which implies by Theorem 3.1.1 that

$$\bar{\mathfrak{A}}[\tau_{\sigma s}^*] = \bar{\mathfrak{A}}[\tau] = \bar{\mathfrak{A}}[\tau_w] = \mathfrak{A}^T.$$

This completes the proof.

Definition 3.1.6 A T^*-algebra $\mathfrak{A}$ on $\mathcal{H}$ is called a WT^*-algebra if $\mathfrak{A} = \mathfrak{A}^T$, equivalently, $\mathfrak{A} = \bar{\mathfrak{A}}[\tau]$, where τ is any of the topologies τ_w, $\tau_{\sigma w}$, τ_s, $\tau_{\sigma s}$, τ_s^* and $\tau_{\sigma s}^*$.

It is clear that if $\mathfrak{A}$ is a WT^*-algebra, then it is a CT^*-algebra.

3.2 Kaplansky Type Density Theorem

In this section we prove the Kaplansky type density theorem using the functional calculus theorem in Sect. 2.3. For $x = (x_n) \in \mathcal{H}_{\mathbb{N}} := \bigoplus_{n=1}^{\infty} \mathcal{H}_n$ (where $\mathcal{H}_n = \mathcal{H}$, $n \in \mathbb{N}$), we define a seminorm p_x on $T^*(\mathcal{H})$ by

$$p_x(A) = \left[\sum_{n=1}^{\infty} \|\pi(A)x_k\|^2 + \sum_{n=1}^{\infty} \|\pi(A)^* x_k\|^2 + \|\lambda(A)\|^2 + \|\lambda(A^\sharp)\|^2 \right]^{\frac{1}{2}}, A \in T^*(\mathcal{H}).$$

As stated in Sect. 2.5, the σ-strong* topology $\tau_{\sigma s}^*$ on $T^*(\mathcal{H})$ is defined by a family $\{p_x;\ x \in \mathcal{H}_{\mathbb{N}}\}$ of seminorms, and $T^*(\mathcal{H})[\tau_{\sigma s}^*]$ is a complete locally convex $*$-algebra. For $\mathfrak{A} \subset T^*(\mathcal{H})$ we denote by $\bar{\mathfrak{A}}[\tau_{\sigma s}^*]$ the σ-strong* closure of $\mathfrak{A}$, and denote by $u(\mathfrak{A})$ and $u(\mathfrak{A})^0$ the unit balls of $\mathfrak{A}$, that is,

$$u(\mathfrak{A}) := \{A \in \mathfrak{A};\quad \|A\| \leqq 1\},$$
$$u(\mathfrak{A})^0 := \{A \in \mathfrak{A};\quad \|A\| < 1\}.$$

The purpose of this section is to prove the following result called Kaplansky type density theorem.

Theorem 3.2.1 (Kaplansky Type Density Theorem) *Let $\mathfrak{A}$ be a T^*-algebra on $\mathcal{H}$. Then*

$$\overline{u(\mathfrak{A})}[\tau_{\sigma s}^*] = u(\bar{\mathfrak{A}}[\tau_{\sigma s}^*]).$$

For the proof of this theorem we may assume without loss of generality that $\mathfrak{A}$ is a CT^*-algebra because $\overline{u(\mathfrak{A})}[\tau_u] = u(\bar{\mathfrak{A}}[\tau_u])$. We first verify the following

Lemma 3.2.2 *Put*

$$u_\pi(\mathfrak{A}) := \{A \in \mathfrak{A}; \quad \|\pi(A)\| \leqq 1\}.$$

Then

$$\overline{u_\pi(\mathfrak{A})}[\tau^*_{\sigma s}] = u_\pi(\bar{\mathfrak{A}}[\tau^*_{\sigma s}]).$$

Proof We consider continuous functions f and g defined by

$$f(t) = \frac{2}{1+t}, \quad t \geqq 0,$$
$$g(t) = \frac{1}{1+\sqrt{1-t}}, \quad 0 \leqq t \leqq 1.$$

Then it follows from Theorem 2.3.11 that for any $A \in T^*(\mathcal{H})$

$$\begin{aligned} F(A) := f(AA^\sharp)A &= \big(\pi(F(A)), \lambda(F(A)), \lambda(F(A)^\sharp)^*\big) \\ &\in u_\pi(CT^*(A)), \end{aligned} \tag{3.2.1}$$

where

$$\pi(F(A)) = 2(I + \pi(A)\pi(A)^*)^{-1}\pi(A),$$
$$\lambda(F(A)) = 2(I + \pi(A)\pi(A)^*)^{-1}\lambda(A),$$
$$\lambda(F(A)^\sharp) = 2(I + \pi(A)^*\pi(A))^{-1}\lambda(A^\sharp),$$

and that for any $A \in u_\pi(T^*(\mathcal{H}))$,

$$\begin{aligned} G(A) := g(AA^\sharp)A &= \big(\pi(G(A)), \lambda(G(A)), \lambda(G(A)^\sharp)^*\big) \\ &\in u_\pi(CT^*(A)), \end{aligned} \tag{3.2.2}$$

where

$$\pi(G(A)) = \Big[I + (I - \pi(A)\pi(A)^*)^{\frac{1}{2}}\Big]^{-1} \pi(A),$$
$$\lambda(G(A)) = \Big[I + (I - \pi(A)\pi(A)^*)^{\frac{1}{2}}\Big]^{-1} \lambda(A)$$
$$\lambda(G(A)^\sharp) = \Big[I + (I - \pi(A)^*\pi(A))^{\frac{1}{2}}\Big]^{-1} \lambda(A^\sharp).$$

Let $A \in u_\pi(T^*(\mathcal{H}))$. Since

$$\begin{aligned}
&\left(I + \pi(G(A))\pi(G(A))^*\right)^{-1} \\
&= \left(I + \left[I + (I - \pi(A)\pi(A)^*)^{\frac{1}{2}}\right]^{-1} \pi(A)\pi(A)^* \left[I + (I - \pi(A)\pi(A)^*)^{\frac{1}{2}}\right]^{-1}\right)^{-1} \\
&= \frac{1}{2}\left[I + (I - \pi(A)\pi(A)^*)^{\frac{1}{2}}\right],
\end{aligned}$$

it follows that

$$\begin{aligned}
\pi(F(G(A))) &= 2\left(I + \pi(G(A))\pi(G(A))^*\right)^{-1} \pi(G(A)) \\
&= \left[I + (I - \pi(A)\pi(A)^*)^{\frac{1}{2}}\right]\left[I + (I - \pi(A)\pi(A)^*)^{\frac{1}{2}}\right]^{-1} \pi(A) \\
&= \pi(A),
\end{aligned}$$

and that

$$\begin{aligned}
\lambda(F(G(A))) &= 2\left(I + \pi(G(A))\pi(G(A))^*\right)^{-1} \lambda(G(A)) \\
&= \left[I + (I - \pi(A)\pi(A)^*)^{\frac{1}{2}}\right]\left[I + (I - \pi(A)\pi(A)^*)^{\frac{1}{2}}\right]^{-1} \lambda(A) \\
&= \lambda(A).
\end{aligned}$$

Furthermore, since

$$\begin{aligned}
\left(I + \pi(G(A))^*\pi(G(A))\right)^{-1} &= \left(I + \pi(A)^*\left[I + (I - \pi(A)^*\pi(A))^{\frac{1}{2}}\right]^{-2} \pi(A)\right)^{-1} \\
&= \left(I + \pi(A)^*\pi(A)\left[I + (I - \pi(A)^*\pi(A))^{\frac{1}{2}}\right]^{-2}\right)^{-1} \\
&= \frac{1}{2}\left[I + (I - \pi(A)^*\pi(A))^{\frac{1}{2}}\right],
\end{aligned}$$

it follows that

$$\begin{aligned}
\lambda(F(G(A)^\sharp)) &= \left[I + (I - \pi(A)^*\pi(A))^{\frac{1}{2}}\right]\left[I + (I - \pi(A)^*\pi(A))^{\frac{1}{2}}\right]^{-1} \lambda(A^\sharp) \\
&= \lambda(A^\sharp).
\end{aligned}$$

Thus we have

$$F(G(A)) = A. \tag{3.2.3}$$

By (3.2.1)–(3.2.3) we have

$$F(\mathfrak{B}) = u_\pi(\mathfrak{B})$$

for every CT^*-algebra $\mathfrak{B}$ on $\mathcal{H}$. Since $\mathfrak{A}$ is a CT^*-algebra by assumption and $\bar{\mathfrak{A}}([\tau^*_{\sigma s}])$ is a CT^*-algebra too, we have

$$F(\mathfrak{A}) = u_\pi(\mathfrak{A}) \tag{3.2.4}$$

and

$$F(\bar{\mathfrak{A}}[\tau^*_{\sigma s}]) = u_\pi(\bar{\mathfrak{A}}[\tau^*_{\sigma s}]). \tag{3.2.5}$$

Furthermore, the map

$$F: \quad A \in T^*(\mathcal{H}) \longmapsto F(A) \in u_\pi(T^*(\mathcal{H})) \tag{3.2.6}$$

is $\tau^*_{\sigma s}$-continuous. Indeed, for any $A, B \in T^*(\mathcal{H})$ we have

$$\begin{aligned}
&\frac{1}{2}(\pi(F(A)) - \pi(F(B)))\\
&\quad = \left(I + \pi(A)\pi(A)^*\right)^{-1}\pi(A) - \left(I + \pi(B)\pi(B)^*\right)^{-1}\pi(B)\\
&\quad = \left(I + \pi(A)\pi(A)^*\right)^{-1}\pi(A) - \pi(B)\left(I + \pi(B)^*\pi(B)\right)^{-1}\\
&\quad = \left(I + \pi(A)\pi(A)^*\right)^{-1}\\
&\qquad \left[(\pi(A) - \pi(B)) + \pi(A)(\pi(B)^* - \pi(A)^*)\pi(B)\right]\left(I + \pi(B)^*\pi(B)\right)^{-1},
\end{aligned}$$

which yields that for any $x = (x_n) \in \mathcal{H}_{\mathbb{N}}$

$$\begin{aligned}
&\sum_{n=1}^{\infty} \|\pi(F(A))x_n - \pi(F(B))x_n\|^2\\
&\quad \leqq 4\left[\sum_{n=1}^{\infty} \|\left(I + \pi(A)\pi(A)^*\right)^{-1}(\pi(A) - \pi(B))\left(I + \pi(B)^*\pi(B)\right)^{-1} x_n\right.\\
&\qquad \left. + \left(I+\pi(A)\pi(A)^*\right)^{-1}\pi(A)(\pi(B)^*-\pi(A)^*)\pi(B)\left(I+\pi(B)^*\pi(B)\right)^{-1} x_n\|^2\right]\\
&\quad = 4\left[\sum_{n=1}^{\infty} \|\pi(A)u_n - \pi(B)u_n\|^2 + \sum_{n=1}^{\infty} \|\pi(B)^* v_n - \pi(A)^* v_n\|^2\right],
\end{aligned}$$

where

$$u_n := (I + \pi(B)^*\pi(B))^{-1} x_n,$$
$$v_n := \pi(B)\left(I + \pi(B)^*\pi(B)\right)^{-1} x_n$$

and $\{u_n\}, \{v_n\} \in \mathcal{H}_{\mathbb{N}}$. Furthermore, we can verify that

$$\begin{aligned}
&\frac{1}{2}\|\lambda(F(A)) - \lambda(F(B))\| \\
&\quad = \|(I + \pi(A)\pi(A)^*)^{-1}\lambda(A) - (I + \pi(B)\pi(B)^*)^{-1}\lambda(B)\| \\
&\quad = \|(I + \pi(A)\pi(A)^*)^{-1}(\lambda(A) - \lambda(B)) \\
&\qquad + \left[(I + \pi(A)\pi(A)^*)^{-1} - (I + \pi(B)\pi(B)^*)^{-1}\right]\lambda(B)\| \\
&\quad \leqq \|\lambda(A) - \lambda(B)\| + \|(I + \pi(A)\pi(A)^*)^{-1}(\pi(B)\pi(B)^* \\
&\qquad - \pi(A)\pi(A)^*)(I + \pi(B)\pi(B)^*)^{-1}\lambda(B)\| \\
&\quad \leqq \|\lambda(A) - \lambda(B)\| + \|(I + \pi(A)\pi(A^*))^{-1}\pi(A)(\pi(B)^* \\
&\qquad - \pi(A)^*)(I + \pi(B)\pi(B)^*)^{-1}\lambda(B)\| \\
&\qquad + \|(I + \pi(A)\pi(A)^*)^{-1}(\pi(B) - \pi(A))\pi(B)^*(I + \pi(B)\pi(B)^*)^{-1}\lambda(B)\| \\
&\quad \leqq \|\lambda(A) - \lambda(B)\| + \|(\pi(A)^* - \pi(B)^*)(I + \pi(B)\pi(B)^*)^{-1}\lambda(B)\| \\
&\qquad + \|(\pi(A) - \pi(B))\pi(B)^*(I + \pi(B)\pi(B)^*)^{-1}\lambda(B)\|,
\end{aligned}$$

and similarly,

$$\begin{aligned}
\frac{1}{2}\|\lambda(F(A)^\sharp) - \lambda(F(B^\sharp))\| \leqq \|\lambda(A^\sharp) - \lambda(B^\sharp)\| \\
+\|(\pi(A) - \pi(B))(I + \pi(B)^*\pi(B))^{-1}\lambda(B^\sharp)\| \\
+\|(\pi(B)^* - \pi(A)^*)\pi(B)(I + \pi(B)^*\pi(B))^{-1}\lambda(B^\sharp)\|.
\end{aligned}$$

Hence F is $\tau_{\sigma s}^*$-continuous. Using these facts, we can prove that

$$\overline{u_\pi(\mathfrak{A})}[\tau_{\sigma s}^*] = u_\pi(\bar{\mathfrak{A}}[\tau_{\sigma s}^*]). \tag{3.2.7}$$

Indeed, it is clear that $u_\pi(\bar{\mathfrak{A}}[\tau_{\sigma s}^*])$ is $\tau_{\sigma s}^*$-closed. Hence,

$$\overline{u_\pi(\mathfrak{A})}[\tau_{\sigma s}^*] \subset u_\pi(\bar{\mathfrak{A}}[\tau_{\sigma s}^*]).$$

Conversely, take an arbitrary $A \in u_\pi(\bar{\mathfrak{A}}[\tau_{\sigma s}^*])$. By (3.2.5), $A = F(B)$ for some $B \in \bar{\mathfrak{A}}[\tau_{\sigma s}^*]$. Then we can take a net $\{B_\alpha\}$ in $\mathfrak{A}$ such that $\tau_{\sigma s}^*\text{-}\lim_\alpha B_\alpha = B$, so that by (3.2.6)

$$\tau_{\sigma s}^* - \lim_\alpha F(B_\alpha) = F(B) = A.$$

Since $F(B_\alpha) \in u_\pi(\mathfrak{A})$ by (3.2.4), we have $A \in \overline{u_\pi(\mathfrak{A})}[\tau_{\sigma s}^*]$. Thus (3.2.7) holds. This complete the proof.

The Proof of Theorem 3.2.1 Put

$$u_\pi(\mathfrak{A})^0 = \{A \in \mathfrak{A};\quad \|\pi(A)\| < 1\}.$$

Take an arbitrary $A \in u_\pi(\mathfrak{A})$. Since

$$\varepsilon A = (\varepsilon\pi(A), \varepsilon\lambda(A), \varepsilon\lambda(A^\sharp)^*) \in u_\pi(\mathfrak{A})^0$$

for each $\varepsilon \in (0, 1)$ and $\tau_{\sigma s}^*\text{-}\lim_{\varepsilon\to 0} \varepsilon A = A$, we have $A \in \overline{u_\pi(\mathfrak{A})^0}[\tau_{\sigma s}^*]$. Therefore

$$\overline{u_\pi(\mathfrak{A})^0}[\tau_{\sigma s}^*] = \overline{u_\pi(\mathfrak{A})}[\tau_{\sigma s}^*]. \tag{3.2.8}$$

Similarly, we can verify that

$$\overline{u(\mathfrak{A})^0}[\tau_{\sigma s}^*] = \overline{u(\mathfrak{A})}[\tau_{\sigma s}^*]. \tag{3.2.9}$$

Take an arbitrary $A \in u(\bar{\mathfrak{A}}[\tau_{\sigma s}^*])^0$. Then, since

$$A \in u_\pi(\bar{\mathfrak{A}}[\tau_{\sigma s}^*]) = \overline{u_\pi(\mathfrak{A})}[\tau_{\sigma s}^*] = \overline{u_\pi(\mathfrak{A})^0}[\tau_{\sigma s}^*]$$

by Lemma 3.2.2 and (3.2.8), it follows that for any $x = (x_n) \in \mathcal{H}_{\mathbb{N}}$ and $0 < \varepsilon < \min(1 - \|\lambda(A)\|, 1 - \|\lambda(A^\sharp)\|)$ there exists an element B of $u_\pi(\mathfrak{A})^0$ such that

$$\begin{aligned} p_x(A - B) &= \Bigg[\sum_{n=1}^{\infty} \|\pi(A)x_n - \pi(B)x_n\|^2 + \sum_{n=1}^{\infty} \|\pi(A)^* x_n - \pi(B)^* x_n\|^2 \\ &\quad + \|\lambda(A) - \lambda(B)\|^2 + \|\lambda(A^\sharp) - \lambda(B^\sharp)\|^2\Bigg]^{\frac{1}{2}} \\ &< \varepsilon. \end{aligned}$$

Hence we get

$$\begin{aligned} &\|\pi(B)\| < 1, \\ &\|\lambda(B)\| < \|\lambda(A)\| + \varepsilon < 1, \\ &\|\lambda(B^\sharp)\| < \|\lambda(A^\sharp)\| + \varepsilon < 1, \end{aligned}$$

so that $B \in u(\mathfrak{A})^0$ and $A \in \overline{u(\mathfrak{A})^0}[\tau^*_{\sigma s}]$. Thus we have

$$u(\bar{\mathfrak{A}}[\tau^*_{\sigma s}]) \subset \overline{u(\mathfrak{A})^0}[\tau^*_{\sigma s}],$$

which yields by (3.2.9) that

$$u(\bar{\mathfrak{A}}[\tau^*_{\sigma s}]) = \overline{u(\bar{\mathfrak{A}}[\tau^*_{\sigma s}])^0}[\tau^*_{\sigma s}] \subset \overline{u(\mathfrak{A})^0}[\tau^*_{\sigma s}] = \overline{u(\mathfrak{A})}[\tau^*_{\sigma s}].$$

The converse inclusion is trivial. This completes the proof.

Corollary 3.2.3 *Let $\mathfrak{A}$ be a T^*-algebra on $\mathcal{H}$. Then*

$$\overline{u(\mathfrak{A})}[\tau] = u(\bar{\mathfrak{A}}[\tau]) = u(\bar{\mathfrak{A}}[\tau^*_{\sigma s}]) = \overline{u(\mathfrak{A})}[\tau^*_{\sigma s}],$$

where τ is any of the topologies τ_w, $\tau_{\sigma w}$, τ_s, $\tau_{\sigma s}$ and τ^*_s. In particular, if $\mathfrak{A}$ is a WT^*-algebra, then the unit ball $u(\mathfrak{A})$ is closed under these topologies.

Proof It is easily shown that $u(\bar{\mathfrak{A}}[\tau_w])$ is weakly closed. Hence it follows from Corollary 3.1.5 that

$$\overline{u(\mathfrak{A})}[\tau^*_{\sigma s}] \subset \overline{u(\mathfrak{A})}[\tau] \subset \overline{u(\mathfrak{A})}[\tau_w] \subset u(\bar{\mathfrak{A}}[\tau_w]) = u(\bar{\mathfrak{A}}[\tau^*_{\sigma s}]),$$

which completes the proof.

For a closed $*$-ideal $N(\mathfrak{A})$ of a CT^*-algebra $\mathfrak{A}$ defined by $N(\mathfrak{A}) = \{A \in \mathfrak{A};\ \pi(A) = 0\}$, we have the following

Corollary 3.2.4 *Let $\mathfrak{A}$ be a CT^*-algebra on $\mathcal{H}$. Then*

$$N(\bar{\mathfrak{A}}[\tau^*_{\sigma s}]) = N(\mathfrak{A}).$$

Proof Take an arbitrary $A \in N(\bar{\mathfrak{A}}[\tau^*_{\sigma s}])$. By Lemma 3.2.2,

$$\begin{aligned} A \in u^\varepsilon_\pi(\bar{\mathfrak{A}}[\tau^*_{\sigma s}]) &:= \{A \in \bar{\mathfrak{A}}[\tau^*_{\sigma s}];\quad \|\pi(A)\| \leqq \varepsilon\} \\ &= \overline{u^\varepsilon_\pi(\mathfrak{A})}[\tau^*_{\sigma s}] \end{aligned}$$

for any $\varepsilon > 0$. Hence, for any $x \in \mathcal{H}_{\mathbb{N}}$ there exists an element B of $u^\varepsilon_\pi(\mathfrak{A})$ such that $p_x(A - B) < \varepsilon$, which implies that

$$\begin{aligned} \|\pi(B)\| &\leqq \varepsilon, \\ \|\lambda(B) - \lambda(A)\| &< \varepsilon, \\ \|\lambda(B^\sharp) - \lambda(A^\sharp)\| &< \varepsilon, \end{aligned}$$

that is, $\|A - B\| \leqq \varepsilon$. Hence, $A \in \bar{\mathfrak{A}}[\tau_u] = \mathfrak{A}$, which completes the proof.

By Corollary 3.2.4 and Theorem 3.2.1, if $\mathfrak{A}$ is a WT^*-algebra (see Definition 3.1.6), then $N(\mathfrak{A})$ is closed under all of topologies τ_w, $\tau_{\sigma w}$, τ_s, $\tau_{\sigma s}$, τ^*_s and $\tau^*_{\sigma s}$.

Chapter 4
Structure of CT^*-Algebras

Chapter 4 is devoted to structure of CT^*-algebras. In Sect. 4.1 it is proved that every CT^*-algebra $\mathfrak{A}$ is decomposed into $\mathfrak{A} = P(\mathfrak{A}) + Z(\mathfrak{A})$ by the nondegenerate part $P(\mathfrak{A})$ and the nulifier $Z(\mathfrak{A})$ of $\mathfrak{A}$ using the corresponding projections $P_{\mathfrak{A}}$ and $N_{\mathfrak{A}}$, and more decompositions of $\mathfrak{A}$ are obtained decomposing the projections $P_{\mathfrak{A}}$ and $N_{\mathfrak{A}}$. Furthermore, the important equalities: $P_{\mathfrak{A}} = I - N_{\mathfrak{A}^\tau}$ and $P_{\mathfrak{A}^\tau} = I - N_{\mathfrak{A}}$ are verified. In Sect. 4.2 the notions of regular, semifinite, nondegenerate, singular and nilpotent CT^*-algebras are defined and their characterizations are investigated. In Sect. 4.3 it is shown that every commutative semisimple CT^*-algebra is isomorphic to the Banach $*$-algebra $C_0(\Omega) \cap L^2(\Omega, \mu)$ consisting of the commutative C^*-algebra $C_0(\Omega)$ of continuous functions on a locally compact space Ω vanishing at infinity on Ω and the L^2-space $L^2(\Omega, \mu)$ defined by a regular Borel measure μ on Ω. The following subjects are discussed: Sect. 4.4.1 definition of regular, (semisimple, nondegenerate, singular and nilpotent) T^*-algebras. Sect. 4.4.2 construction of $\lambda(\tilde{\mathfrak{A}})$ from $\lambda(\mathfrak{A})$. Sect. 4.4.3 construction of semisimple CT^*-algebras from the C^*-algebra obtained by the uniform closure of $\pi(\mathfrak{A})$. Sect. 4.4.4 the semisimplicity and the singularity of T^*-algebras. 3.4.6 Construction of a natural weight on the positive cone $(\pi(\mathfrak{A})'')_+$ of the von Neumann algebra $\pi(\mathfrak{A})''$. Sect. 4.4.6 the density of a $*$-subalgebra of a T^*-algebra under the strong* topology and under the uniform topology.

A. Inoue, *Tomita's Lectures on Observable Algebras in Hilbert Space*,
Lecture Notes in Mathematics 2285, https://doi.org/10.1007/978-3-030-68893-6_4

4.1 Decomposition of CT^*-Algebras

Let $\mathfrak{A}$ be a CT^*-algebra on $\mathcal{H}$. Write

$$\begin{aligned} P(\mathfrak{A}) &= \text{ the closed linear span of } \{AB;\ A,\ B \in \mathfrak{A}\}, \\ N(\mathfrak{A}) &= \{A \in \mathfrak{A};\ \pi(A) = 0\}, \\ Z(\mathfrak{A}) &= \{A \in \mathfrak{A};\ AX = XA = O \text{ for all } X \in \mathfrak{A}\}. \end{aligned}$$

Then, it is simple to show the following

Proposition 4.1.1 *Suppose that $\mathfrak{A}$ is a CT^*-algebra on $\mathcal{H}$. Then $P(\mathfrak{A})$, $N(\mathfrak{A})$ and $Z(\mathfrak{A})$ are closed $*$-ideals of $\mathfrak{A}$ and $N(\mathfrak{A})$ is identical with the radical $R(\mathfrak{A})$ of $\mathfrak{A}$, which contains $Z(\mathfrak{A})$, where*

$$\begin{aligned} R(\mathfrak{A}) := \{A \in \mathfrak{A};\ (\alpha A + XA) &\text{ has an inverse in the unitization } \mathfrak{A}_{\mathbb{1}} \text{ of } \mathfrak{A} \\ &\text{for all } \alpha \in \mathbb{C} \text{ and } X \in \mathfrak{A}\}. \end{aligned}$$

Remark 4.1.2 The map:

$$A + N(\mathfrak{A}) \in \mathfrak{A}/_{N(\mathfrak{A})} \mapsto \pi(A) \in B(\mathcal{H})$$

is a continuous $*$-isomorphism of the quotient Banach $*$-algebra $\mathfrak{A}/_{N(\mathfrak{A})}$ onto the C^*-normed algebra $\pi(\mathfrak{A})$. For the definition of C^*-normed algebra see C.2 in Appendix. If $\pi(\mathfrak{A})$ is a C^*-algebra, then $\mathfrak{A}/_{N(\mathfrak{A})}$ is $*$-isomorphic to $\pi(\mathfrak{A})$ as Banach $*$-algebras, so that $\mathfrak{A}/_{N(\mathfrak{A})}$ is a C^*-algebra.

In order to prove that every CT^*-algebra $\mathfrak{A}$ on $\mathcal{H}$ is decomposed into $P(\mathfrak{A}) + Z(\mathfrak{A})$, we define the following projections $P_{\mathfrak{A}}$ and $N_{\mathfrak{A}}$ by

$$\begin{aligned} P_{\mathfrak{A}} &= \text{Proj } [\pi(\mathfrak{A})\mathcal{H}], \\ N_{\mathfrak{A}} &= \text{Proj } [\lambda(N(\mathfrak{A}))]. \end{aligned}$$

Then, since $[\pi(\mathfrak{A})\mathcal{H}]$ is both $\pi(\mathfrak{A})$- and $\pi(\mathfrak{A})'$-invariant, and $[\lambda(N(\mathfrak{A}))]$ is $\pi(\mathfrak{A})$-invariant, we have $P_{\mathfrak{A}} \in \pi(\mathfrak{A})' \cap \pi(\mathfrak{A})''$ and $N_{\mathfrak{A}} \in \pi(\mathfrak{A})'$. Furthermore, we get, by Corollary C.3 in Appendix C, the following result used often in this note.

$$\text{There exists a net } \{e_\alpha\} \text{ in } u_\pi(\mathfrak{A}) \text{ such that } \pi(e_\alpha) \text{ converges strongly to } P_{\mathfrak{A}}. \tag{4.1.1}$$

For any projection E in $\pi(\mathfrak{A})'$ we put

$$\begin{aligned} \tau_E(A) &= (\pi(A), E\lambda(A), (E\lambda(A^\sharp))^*), \\ \nu_E(A) &= (0, E\lambda(A), (E\lambda(A^\sharp))^*) \end{aligned}$$

for $A \in \mathfrak{A}$, and

$$\tau_E(\mathfrak{A}) = \{\tau_E(A);\ A \in \mathfrak{A}\},$$
$$\nu_E(\mathfrak{A}) = \{\nu_E(A);\ A \in \mathfrak{A}\}.$$

Then, the following is immediate.

Lemma 4.1.3 *Let $\mathfrak{A}$ be a CT^*-algebra on $\mathcal{H}$. Then*

(1) τ_E is a continuous $*$-homomorphism of the Banach $*$-algebra $\mathfrak{A}$ into the Banach $*$-algebra $T^*(\mathcal{H})$, and $\tau_E(\mathfrak{A})$ is a T^*-algebra on $\mathcal{H}$. If $\tau_E(\mathfrak{A}) \subset \mathfrak{A}$, then $\tau_E(\mathfrak{A})$ is a CT^*-algebra on $\mathcal{H}$;
(2) $\nu_E(\mathfrak{A})$ is a T^*-algebra on $\mathcal{H}$ satisfying $\nu_E(\mathfrak{A})^2 = \{O\}$. If $\nu_E(\mathfrak{A}) \subset \mathfrak{A}$, then $\nu_E(\mathfrak{A})$ is a CT^*-algebra on $\mathcal{H}$.

Under these preparations we get a decomposition theorem of a CT^*-algebra.

Theorem 4.1.4 *Let $\mathfrak{A}$ be a CT^*-algebra on $\mathcal{H}$. Then*

(1) $P(\mathfrak{A}) = \tau_{P_{\mathfrak{A}}}(\mathfrak{A})$;
(2) $Z(\mathfrak{A}) = \nu_{I-P_{\mathfrak{A}}}(\mathfrak{A})$;
(3) $\mathfrak{A} = P(\mathfrak{A}) + Z(\mathfrak{A})$.

Proof

(1) Let $A \in \mathfrak{A}$ and denote by P_A the projection onto the subspace $[\pi(A)\mathcal{H}]$. By Theorem 2.3.12, A is decomposed into $A_s + A_n$, where

$$A_s = (\pi(A), P_A\lambda(A), (P_{A^\sharp}\lambda(A^\sharp))^*) \in P(CT^*(A)),$$
$$A_n = (0, (I - P_A)\lambda(A), ((I - P_{A^\sharp})\lambda(A^\sharp))^*) \in N(CT^*(A)).$$

Since $P_A \leqq P_{\mathfrak{A}}$, we have

$$\tau_{P_{\mathfrak{A}}}(A_s) = (\pi(A), P_A\lambda(A), (P_{A^\sharp}\lambda(A^\sharp))^*)$$
$$= A_s \in P(\mathfrak{A})$$

and

$$\tau_{P_{\mathfrak{A}}}(A_n) = (0, P_{\mathfrak{A}}\lambda(A_n), (P_{\mathfrak{A}}\lambda(A_n^\sharp))^*).$$

Furthermore, since

$$e_\alpha A_n + A_n e_\alpha^\sharp = (0, \pi(e_\alpha)(I - P_A)\lambda(A_n), (\pi(e_\alpha)(I - P_{A^\sharp})\lambda(A_n^\sharp))^*)$$
$$\in P(\mathfrak{A})$$

and by (4.1.1)

$$\begin{aligned}\tau_u\text{-}\lim_{\alpha}(e_\alpha A_n + A_n e_\alpha^\sharp) &= (0, P_{\mathfrak{A}}\lambda(A_n), (P_{\mathfrak{A}}\lambda(A_n^\sharp))^*) \\ &= \tau_{P_{\mathfrak{A}}}(A_n),\end{aligned}$$

where $\{e_\alpha\}$ is a net in $u_\pi(\mathfrak{A})$ in (4.1.1), it follows that $\tau_{P_{\mathfrak{A}}}(A_n) \in P(\mathfrak{A})$, which implies that

$$\tau_{P_{\mathfrak{A}}}(A) = A_s + \tau_{P_{\mathfrak{A}}}(A_n) \in P(\mathfrak{A}).$$

Thus $\tau_{P_{\mathfrak{A}}}(\mathfrak{A}) \subset P(\mathfrak{A})$. Hence it follows from Lemma 4.1.3 that $\tau_{P_{\mathfrak{A}}}(\mathfrak{A})$ is a CT^*-algebra on $\mathcal{H}$, so that the converse inclusion $P(\mathfrak{A}) \subset \tau_{P_{\mathfrak{A}}}(\mathfrak{A})$ is trivial.

(2) Take an arbitrary $A \in Z(\mathfrak{A})$. Then, since $\pi(A) = 0$, $\pi(e_\alpha)\lambda(A) = 0$ and $\pi(e_\alpha)\lambda(A^\sharp) = 0$ for all α, it follows from (4.1.1) that $P_{\mathfrak{A}}\lambda(A) = 0$ and $P_{\mathfrak{A}}\lambda(A^\sharp) = 0$; hence $A = \nu_{I-P_{\mathfrak{A}}}(A)$. Thus $Z(\mathfrak{A}) \subset \nu_{I-P_{\mathfrak{A}}}(\mathfrak{A})$. We show the converse. By (1) we see that

$$\nu_{I-P_{\mathfrak{A}}}(A) = A - \tau_{P_{\mathfrak{A}}}(A) \in \mathfrak{A}$$

for all $A \in \mathfrak{A}$, and that

$$\begin{aligned}B\nu_{I-P_{\mathfrak{A}}}(A) &= (0, \pi(B)(I - P_{\mathfrak{A}})\lambda(A), 0) \\ &= (0, (I - P_{\mathfrak{A}})\pi(B)\lambda(A), 0) \\ &= O,\end{aligned}$$

and that

$$\begin{aligned}\nu_{I-P_{\mathfrak{A}}}(A)B &= (0, 0, (\pi(B)^*(I - P_{\mathfrak{A}})\lambda(A^\sharp))^*) \\ &= O\end{aligned}$$

for all $A,\ B \in \mathfrak{A}$, which implies that $\nu_{I-P_{\mathfrak{A}}}(\mathfrak{A}) \subset Z(\mathfrak{A})$. Thus, (2) is proved.

(3) By (1) and(2) we obtain

$$A = \tau_{P_{\mathfrak{A}}}(A) + \nu_{I-P_{\mathfrak{A}}}(A) \in P(\mathfrak{A}) + Z(\mathfrak{A})$$

for all $A \in \mathfrak{A}$. Hence, $\mathfrak{A} \subset P(\mathfrak{A}) + Z(\mathfrak{A})$. The converse is trivial. Thus, $\mathfrak{A} = P(\mathfrak{A}) + Z(\mathfrak{A})$. This completes the proof.

To consider more detailed decompositions of $\mathfrak{A}$, we define the following projections:

$$\begin{aligned} F_{\mathfrak{A}} &= P_{\mathfrak{A}}(I - N_{\mathfrak{A}}), \\ S_{\mathfrak{A}} &= P_{\mathfrak{A}} N_{\mathfrak{A}}, \\ Z_{\mathfrak{A}} &= (I - P_{\mathfrak{A}}) N_{\mathfrak{A}}, \\ V_{\mathfrak{A}} &= (I - P_{\mathfrak{A}})(I - N_{\mathfrak{A}}). \end{aligned}$$

Then, $F_{\mathfrak{A}}$, $S_{\mathfrak{A}}$, $Z_{\mathfrak{A}}$ and $V_{\mathfrak{A}}$ are projections in $\pi(\mathfrak{A})'$, and

$$\begin{aligned} I &= P_{\mathfrak{A}} + (I - P_{\mathfrak{A}}) \\ &= (F_{\mathfrak{A}} + S_{\mathfrak{A}}) + (Z_{\mathfrak{A}} + V_{\mathfrak{A}}) \\ &= F_{\mathfrak{A}} + N_{\mathfrak{A}} + V_{\mathfrak{A}}. \end{aligned}$$

Lemma 4.1.5

(1) $\tau_{F_{\mathfrak{A}}}(\mathfrak{A})$ *is a closed* $*$*-subalgebra of the* CT^**-algebra* $P(\mathfrak{A})[\tau_u]$, *and* $\tau_{F_{\mathfrak{A}}}$ *is a continuous* $*$*-homomorphism of the* CT^**-algebra* $\mathfrak{A}[\tau_u]$ *onto the* CT^**-algebra* $\tau_{F_{\mathfrak{A}}}(\mathfrak{A})[\tau_u]$ *with kernel* $N(\mathfrak{A})$. *Hence,* $\tau_{F_{\mathfrak{A}}}(\mathfrak{A})$ *is* $*$*-isomorphic to* $\mathfrak{A}/_{N(\mathfrak{A})}$.
(2) $\nu_{S_{\mathfrak{A}}}(\mathfrak{A}) = N(S_{\mathfrak{A}}) := \{(0, x, y^*);\ x,\ y \in S_{\mathfrak{A}}\mathcal{H}\} = P(\mathfrak{A}) \cap N(\mathfrak{A})$.
(3) $\nu_{V_{\mathfrak{A}}}(\mathfrak{A}) = \{O\}$.

Proof (2) Take arbitrary $x,\ y \in S_{\mathfrak{A}}\mathcal{H}$. Since $N_{\mathfrak{A}}x = x$ and $N_{\mathfrak{A}}y = y$, we can find sequences $\{A_n\}$ and $\{B_n\}$ in $N(\mathfrak{A})$ such that $\lim_{n\to\infty} \lambda(A_n) = x$ and $\lim_{n\to\infty} \lambda(B_n) = y$. Then, we get that

$$\begin{aligned} e_\alpha A_n + B_n^\sharp e_\alpha^\sharp &= (0, \pi(e_\alpha)\lambda(A_n), (\pi(e_\alpha)\lambda(B_n))^*) \\ &\in P(\mathfrak{A}) \cap N(\mathfrak{A}) \end{aligned}$$

for all α and $n \in \mathbb{N}$, where $\{e_\alpha\}$ is a net in $\mathfrak{A}$ in (4.1.1) and that

$$\begin{aligned} \tau_u - \lim_{\alpha} \lim_{n\to\infty} (e_\alpha A_n + B_n^\sharp e_\alpha^\sharp) &= (0, P_{\mathfrak{A}}x, (P_{\mathfrak{A}}y)^*) \\ &= (0, x, y^*). \end{aligned}$$

Hence, $A := (0, x, y^*) \in P(\mathfrak{A}) \cap N(\mathfrak{A})$ and $\nu_{S_{\mathfrak{A}}}(A) = A \in \nu_{S_{\mathfrak{A}}}(\mathfrak{A})$. Thus we have

$$N(S_{\mathfrak{A}}) = \nu_{S_{\mathfrak{A}}}(\mathfrak{A})$$

and

$$N(S_{\mathfrak{A}}) \subset P(\mathfrak{A}) \cap N(\mathfrak{A}). \tag{4.1.2}$$

Conversely, suppose $A \in P(\mathfrak{A}) \cap N(\mathfrak{A})$. Then, since $A = (0, N_{\mathfrak{A}}\lambda(A), (N_{\mathfrak{A}}\lambda(A^\sharp))^*) \in P(\mathfrak{A})$, it follows that

$$\begin{aligned} A = \tau_{P_{\mathfrak{A}}}(A) &= (0, P_{\mathfrak{A}}N_{\mathfrak{A}}\lambda(A), (P_{\mathfrak{A}}N_{\mathfrak{A}}\lambda(A^\sharp))^*) \\ &= (0, S_{\mathfrak{A}}\lambda(A), (S_{\mathfrak{A}}\lambda(A^\sharp))^*) \\ &= \nu_{S_{\mathfrak{A}}}(A) \\ &\in N(S_{\mathfrak{A}}), \end{aligned}$$

which implies by (4.1.2) that $N(S_{\mathfrak{A}}) = P(\mathfrak{A}) \cap N(\mathfrak{A})$.
(1) By Theorem 4.1.4 and (2) we see

$$\tau_{F_{\mathfrak{A}}}(A) = \tau_{P_{\mathfrak{A}}}(A) + \nu_{S_{\mathfrak{A}}}(A) \in P(\mathfrak{A})$$

for all $A \in \mathfrak{A}$, which yields by Lemma 4.1.3 that $\tau_{F_{\mathfrak{A}}}(\mathfrak{A})$ is a closed $*$-subalgebra of the CT^*-algebra $P(\mathfrak{A})[\tau_u]$, and $\tau_{F_{\mathfrak{A}}}$ is a continuous $*$-homomorphism of the CT^*-algebra $\mathfrak{A}[\tau_u]$ onto the CT^*-algebra $\tau_{F_{\mathfrak{A}}}(\mathfrak{A})[\tau_u]$. It is easily shown that the kernel of $\tau_{F_{\mathfrak{A}}}$ coincides with $N(\mathfrak{A})$. Hence $\tau_{F_{\mathfrak{A}}}(\mathfrak{A})$ is isomorphic to $\mathfrak{A}/N(\mathfrak{A})$.
(3) By Theorem 2.3.12, A is decomposed into $A = A_s + A_n$, where $A_s \in P(\mathfrak{A})$ and $A_n \in N(\mathfrak{A})$. Therefore

$$\begin{aligned} V_{\mathfrak{A}}\lambda(A) &= (I - P_{\mathfrak{A}})(I - N_{\mathfrak{A}})\lambda(A_s) + (I - P_{\mathfrak{A}})(I - N_{\mathfrak{A}})\lambda(A_n) \\ &= 0. \end{aligned}$$

This completes the proof.

By Theorem 4.1.4 and Lemma 4.1.5 we obtain the following decomposition theorem of a CT^*-algebra which is one of the main results of this note.

Theorem 4.1.6 *Let $\mathfrak{A}$ be a CT^*-algebra on $\mathcal{H}$. Then*

$$\begin{aligned} P(\mathfrak{A}) &= \tau_{F_{\mathfrak{A}}}(\mathfrak{A}) + \nu_{S_{\mathfrak{A}}}(\mathfrak{A}), \\ Z(\mathfrak{A}) &= \nu_{I-P_{\mathfrak{A}}}(\mathfrak{A}) = \nu_{Z_{\mathfrak{A}}}(\mathfrak{A}), \\ N(\mathfrak{A}) &= \nu_{S_{\mathfrak{A}}}(\mathfrak{A}) + \nu_{Z_{\mathfrak{A}}}(\mathfrak{A}) = \nu_{S_{\mathfrak{A}}}(\mathfrak{A}) + Z(\mathfrak{A}), \end{aligned}$$

and

$$\begin{aligned} \mathfrak{A} &= P(\mathfrak{A}) + Z(\mathfrak{A}) \\ &= (\tau_{F_{\mathfrak{A}}}(\mathfrak{A}) + \nu_{S_{\mathfrak{A}}}(\mathfrak{A})) + Z(\mathfrak{A}) \\ &= \tau_{F_{\mathfrak{A}}}(\mathfrak{A}) + N(\mathfrak{A}). \end{aligned}$$

For the relationship among the projections $P_{\mathfrak{A}}$, $N_{\mathfrak{A}}$, $P_{\mathfrak{A}^\tau}$, $N_{\mathfrak{A}^\tau}$, $P_{\mathfrak{A}^{\tau\tau}}$ and $N_{\mathfrak{A}^{\tau\tau}}$ we obtain the following important results:

Theorem 4.1.7 *Let $\mathfrak{A}$ be a CT^*-algebra on $\mathcal{H}$. Then*

(1) $P_{\mathfrak{A}^{\tau\tau}} = P_{\mathfrak{A}}$ and $N_{\mathfrak{A}^{\tau\tau}} = N_{\mathfrak{A}}$;
(2) $P_{\mathfrak{A}} = I - N_{\mathfrak{A}^\tau}$ and $P_{\mathfrak{A}^\tau} = I - N_{\mathfrak{A}}$.

To prove Theorem 4.1.7 we prepare some statements.

Lemma 4.1.8 *Suppose $K \in N(T^*(\mathcal{H}))$. Then,*

$$K \in \mathfrak{A}^\tau \quad \textit{if and only if} \quad P_{\mathfrak{A}}\lambda(K) = P_{\mathfrak{A}}\lambda(K^\sharp) = 0.$$

Proof It is easily shown that

$$K \in \mathfrak{A}^\tau \quad \text{if and only if} \quad \pi(A)\lambda(K) = \pi(A)\lambda(K^\sharp) = 0 \quad \text{for all } A \in \mathfrak{A}. \tag{4.1.3}$$

Suppose that $K \in \mathfrak{A}^\tau$. Then it follows from (4.1.1) and (4.1.3) that

$$P_{\mathfrak{A}}\lambda(K) = \lim_\alpha \pi(e_\alpha)\lambda(K) = 0,$$
$$P_{\mathfrak{A}}\lambda(K^\sharp) = \lim_\alpha \pi(e_\alpha)\lambda(K^\sharp) = 0.$$

Conversely, suppose $P_{\mathfrak{A}}\lambda(K) = P_{\mathfrak{A}}\lambda(K^\sharp) = 0$. Then it follows that

$$\pi(A)\lambda(K) = P_{\mathfrak{A}}\pi(A)\lambda(K) = \pi(A)P_{\mathfrak{A}}\lambda(K) = 0,$$
$$\pi(A)\lambda(K^\sharp) = \pi(A)P_{\mathfrak{A}}\lambda(K^\sharp) = 0$$

for all $A \in \mathfrak{A}$, which implies by (4.1.3) that $K \in \mathfrak{A}^\tau$. This completes the proof.

Lemma 4.1.9 $N_{\mathfrak{A}^\tau} = I - P_{\mathfrak{A}}$.

Proof By Lemma 4.1.8 we have

$$(I - P_{\mathfrak{A}})\lambda(K) = \lambda(K) = N_{\mathfrak{A}^\tau}\lambda(K)$$

for all $K \in N(\mathfrak{A}^\tau)$. Hence it follows that $I - P_{\mathfrak{A}} \geqq N_{\mathfrak{A}^\tau}$. Conversely, take an arbitrary $x \in \mathcal{H}$ such that $(I - P_{\mathfrak{A}})x = x$. Then, by Lemma 4.1.8, $\tilde{x} := (0, x, 0) \in N(\mathfrak{A}^\tau)$. Hence, $N_{\mathfrak{A}^\tau}x = x$, so $I - P_{\mathfrak{A}} \leqq N_{\mathfrak{A}^\tau}$. This completes the proof.

Lemma 4.1.10 $P_{\mathfrak{A}^{\tau\tau}} = P_{\mathfrak{A}}$.

Proof By Lemma 4.1.9, we get

$$I - P_{\mathfrak{A}^{\tau\tau}} = N_{\mathfrak{A}^{\tau\tau\tau}} = N_{\mathfrak{A}^\tau} = I - P_{\mathfrak{A}},$$

which completes the proof.

Lemma 4.1.11 $S_{\mathfrak{A}^{\tau\tau}} = S_{\mathfrak{A}}$.

Proof Let $x \in S_{\mathfrak{A}^{\tau\tau}}\mathcal{H}$ and write $\tilde{x} = (0, x, 0)$. By Lemma 4.1.10 we get

$$P_{\mathfrak{A}}x = x. \tag{4.1.4}$$

It is clear that $\tilde{x} \in \mathfrak{A}^{\pi\pi}$. By Lemma 4.1.5,

$$\tilde{x} \in N(S_{\mathfrak{A}^{\tau\tau}}) = P(\mathfrak{A}^{\tau\tau}) \cap N(\mathfrak{A}^{\tau\tau}) \subset \mathfrak{A}^{\tau\tau}.$$

Hence, $K\tilde{x} = \tilde{x}K$ for all $K \in \mathfrak{A}^{\rho}$ ($\subset \mathfrak{A}^{\tau}$). Furthermore, for any $K \in \mathfrak{A}^{\rho}$ we can show that

$$(\lambda(K)|\lambda(\tilde{x}^{\sharp})) = 0,$$

and that by (4.1.1),

$$\begin{aligned}
&\lim_{\alpha}(\pi(\tilde{x})\lambda(K^{\sharp})|\lambda(e_{\alpha}^{\sharp})) \\
&= \lim_{\alpha}(x|\pi(K)^{*}\lambda(e_{\alpha}^{\sharp})) \\
&= \lim_{\alpha}(\pi(e_{\alpha})x|\lambda(K^{\sharp})) \\
&= (x|\lambda(K^{\sharp})) \\
&= (\lambda(\tilde{x})|\lambda(K^{\sharp})) \\
&= 0.
\end{aligned}$$

Hence, $\tilde{x} \in \mathfrak{A}^{\rho\rho}$. Thus $\tilde{x} \in \mathfrak{A}^{\pi\pi} \cap \mathfrak{A}^{\tau\tau} \cap \mathfrak{A}^{\rho\rho} = \mathfrak{A}^{T}$. This implies by the von Neumann type density theorem (Theorem 3.1.1) and Corollary 3.2.4 that

$$\tilde{x} \in N(\mathfrak{A}^{T}) = N(\bar{\mathfrak{A}}[\tau_{\sigma s}^{*}]) = N(\mathfrak{A}).$$

Hence it follows from (4.1.4) that $S_{\mathfrak{A}}x = x$. Therefore $S_{\mathfrak{A}^{\tau\tau}} = S_{\mathfrak{A}}$.

The Proof of Theorem 4.1.7 By Lemma 4.1.10 we get that

$$P_{\mathfrak{A}^{\tau\tau}} = P_{\mathfrak{A}}, \tag{4.1.5}$$

and that by Lemmas 4.1.10 and 4.1.11,

$$\begin{aligned}
P_{\mathfrak{A}}(N_{\mathfrak{A}^{\tau\tau}} - N_{\mathfrak{A}}) &= P_{\mathfrak{A}^{\tau\tau}}N_{\mathfrak{A}^{\tau\tau}} - P_{\mathfrak{A}}N_{\mathfrak{A}} \\
&= S_{\mathfrak{A}^{\tau\tau}} - S_{\mathfrak{A}} \\
&= 0.
\end{aligned} \tag{4.1.6}$$

By (4.1.3), $\tilde{V}_{\mathfrak{A}} := (V_{\mathfrak{A}}, 0, 0) \in \mathfrak{A}^{\rho}$; hence we have

$$\begin{aligned}(0, V_{\mathfrak{A}}\lambda(A), 0) &= \tilde{V}_{\mathfrak{A}} A \\ &= A\tilde{V}_{\mathfrak{A}} \\ &= (0, 0, (V_{\mathfrak{A}}^*\lambda(A^{\sharp}))^*)\end{aligned}$$

for all $A \in \mathfrak{A}^{\tau\tau}$. Therefore $V_{\mathfrak{A}}\lambda(A) = 0$ for all $A \in \mathfrak{A}^{\tau\tau}$, which implies that

$$(I - P_{\mathfrak{A}})(N_{\mathfrak{A}^{\tau\tau}} - N_{\mathfrak{A}}) = N_{\mathfrak{A}^{\tau\tau}} V_{\mathfrak{A}} = 0. \tag{4.1.7}$$

Thus we get $V_{\mathfrak{A}} N_{\mathfrak{A}^{\tau\tau}} = 0$. By (4.1.6) and (4.1.7) we have

$$N_{\mathfrak{A}^{\tau\tau}} = N_{\mathfrak{A}}. \tag{4.1.8}$$

Thus statement (1) holds by (4.1.5) and (4.1.8). By Lemma 4.1.9 we have

$$P_{\mathfrak{A}} = I - N_{\mathfrak{A}^{\tau}},$$

which implies by (4.1.8) that

$$P_{\mathfrak{A}^{\tau}} = I - N_{\mathfrak{A}^{\tau\tau}} = I - N_{\mathfrak{A}}.$$

Hence, statement (2) holds. This completes the proof.

Theorem 4.1.7 shows the following

Corollary 4.1.12 *Let $\mathfrak{A}$ be a CT^*-algebra on $\mathcal{H}$. Then*

$$F_{\mathfrak{A}^{\tau}} = F_{\mathfrak{A}}, \quad S_{\mathfrak{A}^{\tau}} = V_{\mathfrak{A}}, \quad Z_{\mathfrak{A}^{\tau}} = Z_{\mathfrak{A}}, \quad V_{\mathfrak{A}^{\tau}} = S_{\mathfrak{A}}.$$

4.2 Classification of CT^*-Algebras

In this section we define the notions of regular, semifinite, nondegenerate, singular and nilpotent CT^*-algebras and characterize them.

Definition 4.2.1 Let $\mathfrak{A}$ be a CT^*-algebra on $\mathcal{H}$. If $\mathfrak{A} = F(\mathfrak{A}) := \tau_{F_{\mathfrak{A}}}(\mathfrak{A})$, then $\mathfrak{A}$ is called semisimple, and if $\mathfrak{A} = P(\mathfrak{A})$, then $\mathfrak{A}$ is called nondegenerate. If $F_{\mathfrak{A}} = I$, then $\mathfrak{A}$ is called regular, and if $N_{\mathfrak{A}} = I$, then $\mathfrak{A}$ is called singular. If $P_{\mathfrak{A}} = 0$, then $\mathfrak{A}$ is called nilpotent.

We first investigate the relationship among regularity, semisimplicity and nondegenetateness of CT^*-algebras.

Proposition 4.2.2 *Let $\mathfrak{A}$ be a CT^*-algebra on $\mathcal{H}$. Consider the following statements:*

(i) $\mathfrak{A}$ *is regular.*
(ii) $\mathfrak{A}$ *is semisimple.*
(iii) $P_{\mathfrak{A}} = I$.
(iv) $\mathfrak{A}$ *is nondegenerate.*
(v) $Z(\mathfrak{A}) = \{O\}$.

Then

$$\begin{array}{ccc} (i) & \Longrightarrow & (ii) \\ \Downarrow & & \Downarrow \\ (iii) \Rightarrow & (iv) & \Leftrightarrow \ (v). \end{array} \tag{4.2.1}$$

Each of the converse implications in (4.2.1) *don't necessarily hold. Suppose that* $\lambda(\mathfrak{A})$ *is dense in* $\mathcal{H}$, *then (i) and (ii) are equivalent, and (iii) and (iv) are equivalent.*

Proof The implications (i)⇒(ii) and (i)⇒(iii) are trivial, and (ii)⇒(iv) and (iii)⇒(iv) follow from Theorem 4.1.4. The equivalence of (iv) and (v) follows from Theorem 4.1.4 and Theorem 4.1.6. We give examples in next Example 4.2.3 that each of the converse implications in (4.2.1) don't hold. It is clear that if $\lambda(\mathfrak{A})$ is dense in $\mathcal{H}$, then (i) and (ii) are equivalent and (iii) and (iv) are equivalent. This completes the proof.

Example 4.2.3

(1) Let $\mathfrak{A}_0$ be a C^*-algebra on $\mathcal{H}$ which is not nondegenerate. We define a CT^*-algebra $\mathfrak{A}$ on $\mathcal{H}$ by

$$\mathfrak{A} = \{(A_0, A_0x, (A_0^*x)^*); \quad A_0 \in \mathfrak{A}_0\}$$

for some $x \neq 0 \in \mathcal{H}$. Since $N(\mathfrak{A}) = \{O\}$, we have $F_{\mathfrak{A}} = P_{\mathfrak{A}}$ and $\tau_{F_{\mathfrak{A}}}(\mathfrak{A}) = \mathfrak{A}$; hence $\mathfrak{A}$ is semisimple. However, since $\pi(\mathfrak{A}) = \mathfrak{A}_0$ is not nondegenerate, $F_{\mathfrak{A}} = P_{\mathfrak{A}} \neq I$, so that $\mathfrak{A}$ is not regular. Thus, (ii) does not lead to (i). Let $\{e_\alpha^0\}$ be an approximate identity for the C^*-algebra $\mathfrak{A}_0$ and write

$$e_\alpha = (e_\alpha^0, e_\alpha^0 x, (e_\alpha^0 x)^*).$$

Then, for any $A := (A_0, A_0x, (A_0^*x)^*) \in \mathfrak{A}$ it follows that $e_\alpha A \in P(\mathfrak{A})$ and $\lim_\alpha e_\alpha A = A$ uniformly, which implies that $\mathfrak{A} = P(\mathfrak{A})$, that is, $\mathfrak{A}$ is nondegenerate. Thus, (iv) does not lead to (iii).

(2) Let $\mathfrak{A} = T^*(\mathcal{H})$. Since $\pi(\mathfrak{A}) = B(\mathcal{H})$ and $\lambda(N(\mathfrak{A})) = \mathcal{H}$, it follows that $P_{\mathfrak{A}} = I$ and $N_{\mathfrak{A}} = I$, so that $F_{\mathfrak{A}} = 0$, which implies that (iii)⇒(i) and (iv)⇒(ii) don't hold.

Definition 4.2.4 For a T^*-algebra $\mathfrak{A}$ on $\mathcal{H}$, the family of subsets $\{g_\alpha\}$ of $\mathcal{H}$ satisfying the following conditions will be denoted by $\mathfrak{M}_{\mathfrak{A}}$:

$$(\pi(A)g_\alpha|g_\beta) = 0 \text{ if } \alpha \neq \beta \text{ and } \sum_\alpha \|\pi(A)g_\alpha\|^2 \leqq \|\lambda(A)\|^2$$

for all $A \in \mathfrak{A}$.

The main result for the structure of CT^*-algebras is Theorem 4.2.6, however it is convenient first to prove the following preliminary important result:

Proposition 4.2.5 *Let $\mathfrak{A}$ be a CT^*-algebra on $\mathcal{H}$. Then $\mathfrak{A}$ is regular if and only if $\mathfrak{A}^\tau$ is regular. If $\mathfrak{A}$ is regular, then there exists an element $\{g_\alpha\}$ of $\mathfrak{M}_{\mathfrak{A}}$ such that*

$$\lambda(A) = \sum_\alpha \pi(A)g_\alpha \tag{4.2.2}$$

for all $A \in \mathfrak{A}$.

Proof It follows from Theorem 4.1.7 that $F_{\mathfrak{A}} = P_{\mathfrak{A}}(I - N_{\mathfrak{A}}) = P_{\mathfrak{A}}P_{\mathfrak{A}^\tau} = F_{\mathfrak{A}^\tau}$, so that $\mathfrak{A}$ is regular if and only if $\mathfrak{A}^\tau$ is regular. Suppose that $\mathfrak{A}$ is regular. Then if $\lambda(\mathfrak{A}) = \{0\}$, then (4.2.2) is trivial. Hence we may assume that $\mathfrak{A}$ is a regular CT^*-algebra with $\lambda(\mathfrak{A}) \neq \{0\}$. Then we shall show (4.2.2) in the following process:

Step 1 We denote by $\mathfrak{P}$ the set of all $E \in \mathfrak{A}^\tau$ satisfying the following three conditions: $\pi(E)$ is a projection, $\lambda(E) \neq 0$ and $\pi(\tau_{I-\pi(E)}(\mathfrak{A})^\tau)$ is nondegenerate. Then $\mathfrak{P}$ is not empty.

Proof For any $A \in \mathfrak{A}$, $T \in \mathfrak{A}^\tau$ and $x \in \mathcal{H}$ we have

$$\begin{aligned}(\lambda(A)|\pi(T)^*x) &= (\pi(T)\lambda(A)|x)\\ &= (\pi(A)\lambda(T)|x)\\ &= (\lambda(T)|\pi(A)^*x).\end{aligned}$$

Since $\lambda(\mathfrak{A}) \neq \{0\}$ by assumption, $\lambda(A) \neq 0$ for some $A \in \mathfrak{A}$. Since $P_{\mathfrak{A}^\tau} = I$ by regularity, there exist $T \in \mathfrak{A}^\tau$ and $x \in \mathcal{H}$ with $(\lambda(A)|\pi(T)^*x) \neq 0$. Hence the above equality implies that $\lambda(T) \neq 0$. Furthermore, since $P_{\mathfrak{A}} = I$, there exists a net $\{e_\alpha\}$ in $\mathfrak{A}$ such that $\{\pi(e_\alpha)\} \subset u(\pi(\mathfrak{A})_+)$ and converges strongly to I as in (4.1.1). Then we have

$$\begin{aligned}\lim_\alpha(\pi(e_\alpha)\lambda(T^\sharp T)|\lambda(e_\alpha)) &= \lim_\alpha(\pi(T^\sharp T)\lambda(e_\alpha)|\lambda(e_\alpha))\\ &= \lim_\alpha \|\pi(e_\alpha)\lambda(T)\|^2\\ &= \|\lambda(T)\|^2\\ &\neq 0,\end{aligned}$$

so that $\lambda(T^\natural T) = \pi(T)^*\lambda(T) \neq 0$ and $\pi(T^\natural T) \neq 0$. Thus there exists an element K of $\mathfrak{A}^\tau$ such that

$$\pi(K) \geqq 0, \ \pi(K) \neq 0 \ \text{ and } \ \lambda(K) \neq 0. \tag{4.2.3}$$

Using the spectral resolution of $\pi(K)$:

$$\pi(K) = \int_0^{\|\pi(K)\|} t\, dE(t),$$

we can choose $t_0 \in (0, \|\pi(K)\|)$ satisfying

$$H_0 := \pi(K)^{-1}(I - E(t_0)) = \int_{t_0}^{\|\pi(K)\|} t^{-1}\, dE(t) \neq 0. \tag{4.2.4}$$

Since $\pi(K) \in \pi(\mathfrak{A})'$, every element $\pi(A)$ of $\pi(\mathfrak{A})$ commutes with the spectral projections $E(t)$ of $\pi(K)$; hence with $I - E(t_0)$ as well as H_0. We here put

$$E = (I - E(t_0), H_0\lambda(K), (H_0\lambda(K^\natural))^*).$$

Then, for any $A \in \mathfrak{A}$

$$\begin{aligned}
AE &= (\pi(A)(I - E(t_0)), \pi(A)H_0\lambda(K), ((I - E(t_0))\lambda(A^\natural))^*) \\
&= ((I - E(t_0))\pi(A), H_0\pi(K)\lambda(A), (H_0\pi(K^\natural)\lambda(A^\natural))^*) \\
&= ((I - E(t_0))\pi(A), (I - E(t_0))\lambda(A), (\pi(A)^* H_0\lambda(K^\natural))^*) \\
&= EA;
\end{aligned}$$

hence, $E \in \mathfrak{A}^\tau$. Furthermore, by (4.2.3) and (4.2.4)

$$E \in \mathfrak{A}^\tau, \ \pi(E) = I - E(t_0) \ \text{ and } \ \lambda(E) = H_0\lambda(K) \neq 0. \tag{4.2.5}$$

Furthermore, we can prove that

$$\pi(\tau_{I-\pi(E)}(\mathfrak{A})^\tau) = \pi(\tau_{E(t_0)}(\mathfrak{A})^\tau) \ \text{ is } \ \text{nondegenerate.} \tag{4.2.6}$$

Indeed, since

$$\begin{aligned}
&(I - E(t_0), 0, 0)(\pi(A), E(t_0)\lambda(A), (E(t_0)\lambda(A^\natural))^*) \\
&= ((I - E(t_0))\pi(A), 0, 0) \\
&= (\pi(A)(I - E(t_0)), 0, 0) \\
&= (\pi(A), E(t_0)\lambda(A), (E(t_0)\lambda(A^\natural))^*)(I - E(t_0), 0, 0)
\end{aligned}$$

and

$$\begin{aligned}
&(\pi(K)E(t_0), E(t_0)\lambda(K), (E(t_0)\lambda(K^\sharp))^*)(\pi(A), E(t_0)\lambda(A), (E(t_0)\lambda(A^\sharp))^*) \\
&= (\pi(K)E(t_0)\pi(A), \pi(K)E(t_0)\lambda(A), (\pi(A)^*E(t_0)\lambda(K^\sharp))^*) \\
&= (\pi(A)\pi(K)E(t_0), \pi(A)E(t_0)\lambda(K), (\pi(K)^*E(t_0)\lambda(A^\sharp))^*) \\
&= (\pi(A), E(t_0)\lambda(A), (E(t_0)\lambda(A^\sharp))^*)(\pi(K)E(t_0), E(t_0)\lambda(K), (E(t_0)\lambda(K^\sharp))^*)
\end{aligned}$$

for all $A \in \mathfrak{A}$, it follows that both $(I - E(t_0), 0, 0)$ and $(\pi(K)E(t_0), E(t_0)\lambda(K), (E(t_0)\lambda(K^\sharp))^*)$ belong to $\tau_{E(t_0)}(\mathfrak{A})^\tau$. Suppose now $y \in (\pi(\tau_{E(t_0)}(\mathfrak{A})^\tau)\mathcal{H})^\perp$. Then, $E(t_0)y = y$ and $(x|\pi(K)y) = (\pi(K)E(t_0)x|y) = 0$ for all $x \in \mathcal{H}$. Hence $\pi(K)y = 0$ and $y = 0$ by the existence of $\pi(K)^{-1}$, which implies that $\pi(\tau_{E(t_0)}(\mathfrak{A})^\tau)$ is nondegenerate. Thus it follows from (4.2.5) and (4.2.6) that $E \in \mathfrak{P}$.

Step 2 Let $\{E_\alpha\}_{\alpha\in\Lambda}$ be a maximal family in $\mathfrak{P}$ such that $\{\pi(E_\alpha)\}_{\alpha\in\Lambda}$ is mutually orthogonal and write $g_\alpha = \lambda(E_\alpha)$ for $\alpha \in \Lambda$. Then $\{g_\alpha\} \in \mathfrak{M}_{\mathfrak{A}}$.

Proof Since $P_{\mathfrak{A}^\tau} = I$, there exists a net $\{e'_\gamma\}$ in $u_\pi(\mathfrak{A}^\tau)$ such that $\pi(e'_\gamma)$ converges strongly to I as in (4.1.1). Then we can show that

$$\begin{aligned}
(\pi(A)g_\alpha|g_\beta) &= (\pi(A)\lambda(E_\alpha)|\lambda(E_\beta)) \\
&= (\lambda(A)|\lambda(E_\alpha E_\beta)) \\
&= \lim_\gamma(\pi(e'_\gamma)\lambda(A)|\lambda(E_\alpha E_\beta)) \\
&= \lim_\gamma(\lambda(e'_\gamma)|\pi(E_\alpha)\pi(E_\beta)\lambda(A^\sharp)) \\
&= 0
\end{aligned}$$

for all $A \in \mathfrak{A}$ and $\alpha \neq \beta$, and furthermore that

$$\sum_{\alpha\in\Lambda} \|\pi(A)g_\alpha\|^2 = \sum_\alpha \|\pi(E_\alpha)\lambda(A)\|^2 \leqq \|\lambda(A)\|^2$$

for all $A \in \mathfrak{A}$. Thus, $\{g_\alpha\} \in \mathfrak{M}_{\mathfrak{A}}$.

Let $E^0_\Lambda = I - \sum_{\alpha\in\Lambda}\pi(E_\alpha)$. Assume now that $\pi(\tau_{E^0_\Lambda}(\mathfrak{A})^\tau)$ is nondegenerate. We will show that $E^0_\Lambda\lambda(A) = 0$, equivalently, $\lambda(A) = \sum_{\alpha\in\Lambda}\pi(A)g_\alpha$ for all $A \in \mathfrak{A}$. After this we assume that there exists an element B of $\mathfrak{A}$ such that

$$E^0_\Lambda\lambda(B) \neq 0. \tag{4.2.7}$$

Step 3 There exists an element S of $u_\pi(\tau_{E^0_\Lambda}(\mathfrak{A})^\tau)$ such that $\pi(S) \neq 0$ and $\pi(B)E^0_\Lambda\lambda(S) \neq 0$.

Proof Since $\pi(\tau_{E^0_\Lambda}(\mathfrak{A})^\tau)$ is nondegenerate, there exists a net $\{S_\alpha\}$ in $u_\pi(\tau_{E^0_\Lambda}(\mathfrak{A})^\tau)$ such that $\pi(S_\alpha)$ converges strongly to I. Then

$$\begin{aligned}\lim_\alpha \pi(B)E^0_\Lambda\lambda(S_\alpha) &= \lim_\alpha E^0_\Lambda\pi(B)\lambda(S_\alpha)\\ &= \lim_\alpha E^0_\Lambda\pi(S_\alpha)E^0_\Lambda\lambda(B)\\ &= E^0_\Lambda\lambda(B)\\ &\neq 0\end{aligned}$$

by assumption (4.2.7). Thus there exists an element $S \in u_\pi(\tau_{E^0_\Lambda}(\mathfrak{A})^\tau)$ with $\pi(S) \neq 0$ and $\pi(B)E^0_\Lambda\lambda(S) \neq 0$.

Step 4 Let $g = E^0_\Lambda\lambda(S)$. Then there exists an element C_0 of $\pi(\mathfrak{A})'$ such that

$$\begin{aligned}C_0E^0_\Lambda\lambda(A) &= \pi(A)g \quad \text{for} \quad \text{all } A \in \mathfrak{A},\\ C_0x &= 0 \quad \text{for} \quad \text{all } x \in (E^0_\Lambda\lambda(\mathfrak{A}))^\perp.\end{aligned} \tag{4.2.8}$$

Proof Since $S \in u_\pi(\tau_{E^0_\Lambda}(\mathfrak{A})^\tau)$, we have

$$\|\pi(A)g\| = \|\pi(A)E^0_\Lambda\lambda(S)\| = \|E^0_\Lambda\pi(S)E^0_\Lambda\lambda(A)\| \leqq \|E^0_\Lambda\lambda(A)\|$$

for all $A \in \mathfrak{A}$. Hence our assertion follows.

Step 5 Let $C_0 = U|C_0|$ be the polar decomposition of C_0 and let $C = (|C_0|, U^*g, (U^*g)^*)$ for $g = E^0_\Lambda\lambda(S)$. Then $C \in \mathfrak{A}^\tau$.

Proof By Step 4, we have

$$\begin{aligned}|C_0|E^0_\Lambda|C_0| = |C_0|E^0_\Lambda &= |C_0|,\\ U^*U &\leqq E^0_\Lambda,\end{aligned} \tag{4.2.9}$$

which implies that

$$\begin{aligned}AC &= (\pi(A)|C_0|, \pi(A)U^*g, (|C_0|\lambda(A^\sharp))^*)\\ &= (|C_0|\pi(A), U^*\pi(A)g, (C_0E^0_\Lambda\lambda(A^\sharp))^*)\\ &= (|C_0|\pi(A), U^*C_0E^0_\Lambda\lambda(A), (\pi(A)^*U^*g)^*)\\ &= CA\end{aligned}$$

for all $A \in \mathfrak{A}$. Hence, $C \in \mathfrak{A}^\tau$.

Step 6 Let $|C_0| = \int_0^{\|C_0\|} \lambda dF(t)$ be the spectral resolution of $|C_0|$. Then there exists $t_0 \in (0, \|C_0\|)$ such that $I - F(t_0) \neq 0$, $I - F(t_0) \leqq E_\Lambda^0$, $F_{t_0} := (I - F(t_0), |C_0|^{-1}(I - F(t_0))\lambda(C), (|C_0|^{-1}(I - F(t_0))\lambda(C))^*) \in \mathfrak{A}^\tau$ and $\lambda(F_{t_0}) \neq 0$.

Proof Since $U|C_0|E_\Lambda^0\lambda(B) = \pi(B)g \neq 0$, it follows from (4.2.8) that $|C_0| \neq 0$ and $U^*g \neq 0$. Hence there exists $t_0 \in (0, \|C_0\|)$ such that

$$I - F(t_0) \neq 0. \tag{4.2.10}$$

Since $C \in \mathfrak{A}^\tau$, it follows that

$$\begin{aligned}
AF_{t_0} &= (\pi(A)(I - F(t_0)), \pi(A)|C_0|^{-1}(I - F(t_0))\lambda(C), ((I - F(t_0))\lambda(A^\sharp))^*) \\
&= ((I - F(t_0))\pi(A), |C_0|^{-1}(I - F(t_0))|C_0|\lambda(A), (|C_0|^{-1}(I - F(t_0))\pi(C)\lambda(A^\sharp))^*) \\
&= ((I - F(t_0))\pi(A), (I - F(t_0))\lambda(A), (\pi(A)^*|C_0|^{-1}(I - F(t_0))\lambda(C))^*) \\
&= F_{t_0}A
\end{aligned}$$

for all $A \in \mathfrak{A}$, which means that

$$F_{t_0} \in \mathfrak{A}^\tau. \tag{4.2.11}$$

Furthermore, we can prove that

$$I - F(t_0) \leqq E_\Lambda^0 \tag{4.2.12}$$

and

$$\lambda(F_{t_0}) = |C_0|^{-1}(I - F(t_0))\lambda(C) \neq 0. \tag{4.2.13}$$

Indeed, since

$$\begin{aligned}
E_\Lambda^0(I - F(t_0)) &= E_\Lambda^0|C_0|(|C_0|^{-2}|C_0|)(I - F(t_0)) \\
&= I - F(t_0)
\end{aligned}$$

by (4.2.9), we have (4.2.12). Suppose now that $\lambda(F_{t_0}) = 0$. Then, it follows from (4.2.12) that

$$(I - F(t_0))E_\Lambda^0\lambda(A) = \pi(F_{t_0})\lambda(A) = \pi(A)\lambda(F_{t_0}) = 0$$

for all $A \in \mathfrak{A}$ and

$$(I - F(t_0))x = |C_0|^{-1}(I - F(t_0))|C_0|x = 0$$

for all $x \in (E^0_\Lambda \lambda(\mathfrak{A}))^\perp$, which implies that $(I - F(t_0)) = 0$. This contradicts (4.2.10). Thus, $\lambda(F_{t_0}) \neq 0$.

Step 7 Let $F^0_\Lambda = I - \left(\sum_{\alpha\in\Lambda} \pi(E_\alpha) + \pi(F_{t_0})\right) = E^0_\Lambda - \pi(F_{t_0})$. Then

$$\pi(\tau_{F^0_\Lambda}(\mathfrak{A})^\tau) \quad \text{is} \quad \text{nondegenerate}. \tag{4.2.14}$$

Proof Take an arbitrary $K \in \tau_{E^0_\Lambda}(\mathfrak{A})^\tau$. Then it follows that

$$\begin{aligned}\pi(K)F^0_\Lambda\lambda(A) &= \pi(K)E^0_\Lambda\lambda(A) - \pi(K)\pi(F_{t_0})\lambda(A)\\ &= \pi(A)(\lambda(K) - \pi(K)\lambda(F_{t_0})),\end{aligned}$$

and that

$$\pi(K)^* F^0_\Lambda\lambda(A) = \pi(A)(\lambda(K^\sharp) - \pi(K)^*\lambda(F_{t_0}))$$

for all $A \in \mathfrak{A}$, which implies that

$$\left(\pi(K), \lambda(K) - \pi(K)\lambda(F_{t_0}), (\lambda(K^\sharp) - \pi(K)^*\lambda(F_{t_0}))^*\right) \in \tau_{F^0_\Lambda}(\mathfrak{A})^\tau.$$

Hence, $\pi(\tau_{E^0_\Lambda}(\mathfrak{A})^\tau) \subset \pi(\tau_{F^0_\Lambda}(\mathfrak{A})^\tau)$, which yields (4.2.14) since $\pi(\tau_{E^0_\Lambda}(\mathfrak{A})^\tau)$ is nondegenerate.

Step 8 $\lambda(A) = \sum_{\alpha\in\Lambda} \pi(A)g_\alpha$ for all $A \in \mathfrak{A}$.

Proof By Step 6, 7, $\{E_\alpha\}\cup\{F_{t_0}\}$ is an element of $\mathfrak{P}$ such that $\{\pi(E_\alpha)\}_{\alpha\in\Lambda}\cup\{\pi(F_{t_0})\}$ are mutually orthogonal and $\tau_{I-(\sum_{\alpha\in\Lambda}\pi(E_\alpha)+\pi(F_{t_0}))}(\mathfrak{A})^\tau$ is nondegenerate. This contradicts to the maximality of $\{E_\alpha\}_{\alpha\in\Lambda}$. Hence, assumption (4.2.7) is not true, that is,

$$E_\Lambda\lambda(A) = 0, \quad \text{equivalently} \quad \lambda(A) = \sum_\alpha \pi(E_\alpha)\lambda(A) = \sum_\alpha \pi(A)g_\alpha$$

for all $A \in \mathfrak{A}$.

This completes the proof of Proposition 4.2.5.

The following theorem for the semisimplicity of CT^*-algebras is one of the most important results of this note:

Theorem 4.2.6 *Let $\mathfrak{A}$ be a CT^*-algebra on $\mathcal{H}$. Then the following statements are equivalent:*

(i) *$\mathfrak{A}$ is semisimple.*
(ii) *$\pi\lceil_{\mathfrak{A}}$ is an injection.*
(iii) *$N(\mathfrak{A}) = \{O\}$.*
(iv) *$\pi(\mathfrak{A}^\tau)$ is nondegenerate.*

(v) *There exists an element* $\{g_\alpha\}$ *of* $\mathfrak{M}_{\mathfrak{A}}$ *such that*

$$\lambda(A) = \sum_\alpha \pi(A) g_\alpha$$

for all $A \in \mathfrak{A}$.

(vi) $\|\lambda(A)\| = \sup\{\|\pi(A)\lambda(K)\|;\quad K \in u_\pi(\mathfrak{A}^\tau)\}$ *for all* $A \in \mathfrak{A}$.

(vii) *The map:* $\pi(A) \to \lambda(A)$ *of* $\pi(\mathfrak{A})[\tau_u^\pi]$ *into the Hilbert space* $\mathcal{H}$ *is closable.*

(viii) *The map:* $\pi(A) \to \lambda(A)$ *of* $\pi(\mathfrak{A})[\tau_{\sigma s^*}^\pi]$ *into the Hilbert space* $\mathcal{H}$ *is closable.*

(viv) *The map:* $\pi(A) \to \nu(A)$ *of* $\pi(\mathfrak{A})[\tau_u^\pi]$ *into the Hilbert space* $N(T^*(\mathcal{H}))$ *is closable.*

(vv) *The map:* $\pi(A) \to \nu(A)$ *of* $\pi(\mathfrak{A})[\tau_{\sigma s^*}^\pi]$ *into the Hilbert space* $N(T^*(\mathcal{H}))$ *is closable.*

In (vii)–(vv) τ_u^π *and* $\tau_{\sigma s^*}^\pi$ *are the uniform topology and the* σs^**-topology on* $\pi(\mathfrak{A})$*, respectively, and the Hilbert space* $\mathcal{H}$ *and* $N(T^*(\mathcal{H}))$ *are equipped with the norm topologies.*

Proof The implications (i)⇒(iii)⇒(ii) are trivial.
(ii)⇒(i) For any $A \in \mathfrak{A}$ we have

$$A - \tau_{F_{\mathfrak{A}}}(A) = (0, \lambda(A) - F_{\mathfrak{A}}\lambda(A), (\lambda(A^\sharp) - F_{\mathfrak{A}}\lambda(A^\sharp))^*),$$

which implies by (ii) that $\lambda(A) = F_{\mathfrak{A}}\lambda(A)$ and $\lambda(A^\sharp) = F_{\mathfrak{A}}\lambda(A^\sharp)$. Hence $\mathfrak{A} = \tau_{F_{\mathfrak{A}}}(\mathfrak{A})$, which means that $\mathfrak{A}$ is semisimple.
The equivalence of (iii) and (iv) follows from Theorem 4.1.7. Thus, (i)–(iv) are equivalent.
(i)⇒(v) By the semisimplicity of $\mathfrak{A}$, we have

$$F_{\mathfrak{A}} = P_{\mathfrak{A}}, \quad \pi(A)P_{\mathfrak{A}} = \pi(A) \quad \text{and} \quad P_{\mathfrak{A}}\lambda(A) = \lambda(A)$$

for all $A \in \mathfrak{A}$, which implies that

$$\mathfrak{A}_{P_{\mathfrak{A}}} := \{(\pi(A)\lceil_{P_{\mathfrak{A}}\mathcal{H}}, \lambda(A), \lambda(A^\sharp)^*);\quad A \in \mathfrak{A}\}$$

is a regular CT^*-algebra on $P_{\mathfrak{A}}\mathcal{H}$. Hence it follows from Proposition 4.2.5 that there exists an element $\{g_\alpha\}$ of $\mathfrak{M}_{\mathfrak{A}_{P_{\mathfrak{A}}}}$ such that

$$\lambda(A) = \sum_\alpha \pi(A) g_\alpha$$

for all $A \in \mathfrak{A}$.Furthermore, it is easily shown that $\mathfrak{M}_{P_{\mathfrak{A}}} \subset \mathfrak{M}_{\mathfrak{A}}$, so that (v) holds.

(v)⇒(ii) This is trivial. Thus, (i)–(v) are equivalent.
(iv)⇒(vi) By assumption (iv) there exists a net $\{e'_\gamma\}$ in $u_\pi(\mathfrak{A}^\tau)$ such that $\pi(e'_\gamma)$ converges strongly to I as in (4.1.1). Then we have

$$\begin{aligned}\lim_\gamma \|\pi(A)\lambda(e'_\gamma)\| &= \lim_\gamma \|\pi(e'_\gamma)\lambda(A)\| \\ &= \|\lambda(A)\|\end{aligned}$$

for all $A \in \mathfrak{A}$, which implies (vi).
(vi)⇒(ii) This is trivial.
(iv)⇒(viii) Suppose that $\{A_\alpha\}$ is any net in $\mathfrak{A}$ such that $\{\pi(A_\alpha)\}$ converges σ-strongly* to 0 and $\{\lambda(A_\alpha)\}$ converges to some element x of $\mathcal{H}$. Since $\pi(\mathfrak{A}^\tau)$ is nondegenerate, it follows that

$$\begin{aligned}x &= \lim_\gamma \pi(e'_\gamma)x \\ &= \lim_\gamma \lim_\alpha \pi(e'_\gamma)\lambda(A_\alpha) \\ &= \lim_\gamma \lim_\alpha \pi(A_\alpha)\lambda(e'_\gamma) \\ &= 0.\end{aligned}$$

Thus, $\pi(A) \to \lambda(A)$ is σ-strongly* closable.
(viii)⇒(vii) This is trivial.
(viii)⇒(vv)⇒(viv) This is trivial.
(vii)⇒(iii) Take an arbitrary $A \in N(\mathfrak{A})$, and put $A_n = A$, $n \in \mathbb{N}$. Then since

$$\pi(A_n) = 0, \quad \lambda(A_n) = \lambda(A) \quad \text{and} \quad \lambda(A_n^\sharp) = \lambda(A^\sharp)$$

for any $n \in \mathbb{N}$, it follows from (vi) that $\lambda(A) = \lambda(A^\sharp) = 0$. Hence, $A = O$. Similarly we can prove (viv)⇒(iii). Thus, (i)–(vv) are equivalent, which completes the proof.

Since $\tau_{F_\mathfrak{A}}(\mathfrak{A})$ is a semisimple CT^*-algebra on $\mathcal{H}$ for every CT^*-algebra $\mathfrak{A}$ on $\mathcal{H}$, Theorem 4.2.6 proves the following

Corollary 4.2.7 *Let $\mathfrak{A}$ be a CT^*-algebra on $\mathcal{H}$. Then there exists an element $\{g_\alpha\}$ of $\mathfrak{M}_{\mathfrak{A}_{F_\mathfrak{A}}}$ such that*

$$F_\mathfrak{A}\lambda(A) = \sum_\alpha \pi(A)g_\alpha$$

for all $A \in \mathfrak{A}$, *and*

$$\begin{aligned}\|F_{\mathfrak{A}}\lambda(A)\|^2 &= \sum_{\alpha} \|\pi(A)g_\alpha\|^2 \\ &= \sup\{\|\pi(A)\lambda(K)\|; \quad K \in u_\pi(\tau_{F_{\mathfrak{A}}}(\mathfrak{A})^\tau)\}\end{aligned}$$

for all $A \in \mathfrak{A}$.

For the commutants and bicommutants of semisimple CT^*-algebra we have the following

Theorem 4.2.8 *Suppose that* $\mathfrak{A}$ *is a semisimple* CT^**-algebra on* $\mathcal{H}$. *Then*

$$\mathfrak{A}^\tau = \mathfrak{A}^\rho, \quad \mathfrak{A}^{\rho\rho} \subset \mathfrak{A}^{\tau\tau} \subset \mathfrak{A}^{\pi\pi} \quad \text{and} \quad \mathfrak{A}^T = \mathfrak{A}^{\rho\rho}.$$

In particular, if $\mathfrak{A}$ *is regular, then*

$$\mathfrak{A}^T = \mathfrak{A}^{\tau\tau} = \mathfrak{A}^{\rho\rho}.$$

Proof For any $S,\ K \in \mathfrak{A}^\tau$ and $C_0 \in \pi(\mathfrak{A})'$ we have

$$\begin{aligned}A(S\tilde{C_0}K) &= \big(\pi(A)\pi(S)C_0\pi(K), \pi(A)\pi(S)C_0\lambda(K), (\pi(K)^*C_0^*\pi(S)^*\lambda(A^\sharp))^*\big) \\ &= \big(\pi(S)C_0\pi(K)\pi(A), \pi(S)C_0\pi(K)\lambda(A), (\pi(A)^*\pi(K)^*C_0^*\lambda(S^\sharp))^*\big) \\ &= (S\tilde{C_0}K)A\end{aligned}$$

for all $A \in \mathfrak{A}$, where $\tilde{C_0} := (C_0, 0, 0)$. Hence $S\tilde{C_0}K \in \mathfrak{A}^\tau$. Furthermore, we get that

$$\begin{aligned}(\lambda(S\tilde{C_0}K)|\lambda(A^\sharp)) &= (\pi(S)C_0\lambda(K)|\lambda(A^\sharp)) \\ &= (C_0\lambda(K)|\pi(A)^*\lambda(S^\sharp)) \\ &= (C_0\pi(K)\lambda(A)|\lambda(S^\sharp)) \\ &= (\lambda(A)|\pi(K)^*C_0^*\lambda(S^\sharp)) \\ &= (\lambda(A)|\lambda((S\tilde{C_0}K)^\sharp))\end{aligned}$$

for all $A \in \mathfrak{A}$. Hence it follows that

$$S\tilde{C_0}K \in \mathfrak{A}^\rho \tag{4.2.15}$$

for all $S,\ K \in \mathfrak{A}^\tau$ and $C_0 \in \pi(\mathfrak{A})'$. We show that

$$\mathfrak{A}^{\tau\tau} \subset \mathfrak{A}^{\pi\pi}. \tag{4.2.16}$$

Indeed, since $\mathfrak{A}^{\pi\pi} = \{A \in T^*(\mathcal{H}); \quad \pi(A) \in \pi(\mathfrak{A})''\}$, it suffices to show that $\pi(A) \in \pi(\mathfrak{A})''$ for all $A \in \mathfrak{A}^{\tau\tau}$. Take an arbitrary $A \in \mathfrak{A}^{\tau\tau}$. Since $\pi(\mathfrak{A}^{\tau})$ is nondegenerate by Theorem 4.2.6, as seen often, there exists a net $\{e'_\gamma\}$ in $u_\pi(\mathfrak{A}^{\tau})$ such that $\pi(e'_\gamma)$ converges strongly to I. By (4.2.15) we have

$$(e'_\gamma)^\sharp \tilde{C}_0 e'_\gamma \in \mathfrak{A}^{\rho} \subset \mathfrak{A}^{\tau}$$

for all γ and $C_0 \in \pi(\mathfrak{A})'$, so that

$$\pi(e'_\gamma)^* C_0 \pi(e'_\gamma)\pi(A) = \pi(A)\pi(e'_\gamma)^* C_0 \pi(e'_\gamma),$$

$$\begin{aligned}\lim_\gamma \left(\pi(e'_\gamma)^* C_0 \pi(e'_\gamma)\pi(A)x|y\right) &= \lim_\gamma \left(C_0\pi(e'_\gamma)\pi(A)x|\pi(e'_\gamma)y\right) \\ &= (C_0\pi(A)x|y)\end{aligned}$$

and

$$\lim_\gamma \left(\pi(A)\pi(e'_\gamma)^* C_0 \pi(e'_\gamma)x|y\right) = (C_0 x|\pi(A)^* y)$$

for all $A \in \mathfrak{A}$ and $x,\ y \in \mathcal{H}$. Hence, $\pi(A) \in \pi(\mathfrak{A})''$. Thus, (4.2.16) holds. Since $P_{\mathfrak{A}^{\tau}} = I$, it follows from Proposition 4.2.2 that $\mathfrak{A}^{\tau} = P(\mathfrak{A}^{\tau})$, which implies by (4.2.15) that $\mathfrak{A}^{\tau} = P(\mathfrak{A}^{\tau}) \subset \mathfrak{A}^{\rho}$. Therefore

$$\mathfrak{A}^{\tau} = \mathfrak{A}^{\rho} \quad \text{and} \quad \mathfrak{A}^{\rho\rho} = \mathfrak{A}^{\tau\rho} \subset \mathfrak{A}^{\tau\tau}. \tag{4.2.17}$$

By (4.2.16) and (4.2.17) we get

$$\mathfrak{A}^{\rho\rho} \subset \mathfrak{A}^{\tau\tau} \subset \mathfrak{A}^{\pi\pi} \quad \text{and} \quad \mathfrak{A}^{T} = \mathfrak{A}^{\rho\rho}.$$

Suppose now that $\mathfrak{A}$ is regular. By Proposition 4.2.5, $\mathfrak{A}^{\tau}$ is regular, so that it follows from (4.2.17) that

$$\mathfrak{A}^{\tau\tau} = \mathfrak{A}^{\tau\rho} = \mathfrak{A}^{\rho\rho}.$$

This completes the proof.

Remark *By Proposition 4.2.5, $\mathfrak{A}$ is regular if and only if $\mathfrak{A}^{\tau}$ is regular, but we don't know the relationship between the semisimplicity of $\mathfrak{A}$ and of $\mathfrak{A}^{\tau}$.*

For the singularity of $\mathfrak{A}$ we have the following

Theorem 4.2.9 *Let $\mathfrak{A}$ be a CT^*-algebra on $\mathcal{H}$. Then the following statements hold:*

(1) $\mathfrak{A}$ is singular if and only if for any $x \in \mathcal{H}$ there exists a sequence $\{A_n\}$ in $N(\mathfrak{A})$ such that $\lim_{n\to\infty} \lambda(A_n) = x$.

(2) *$\mathfrak{A}$ is singular and nondegenerate if and only if $S_{\mathfrak{A}} = I$. In this case, we have*

$$\mathfrak{A}^{\tau} = \{O\} \quad \text{and} \quad \mathfrak{A}^{T} = \mathfrak{A}^{\pi\pi}.$$

(3) *Suppose that $\mathfrak{A}$ is nondegenerate. Then the following (i)–(iv) are equivalent:*

(i) *$\mathfrak{A}$ is singular.*
(ii) *For any $x,\ y \in \mathcal{H}$ there exists an element A of $N(\mathfrak{A})$ such that $\lambda(A) = x$ and $\lambda(A^{\sharp}) = y$.*
(iii) *$\pi(\mathfrak{A})$ is a C^*-algebra on $\mathcal{H}$, and*

$$(\pi(\mathfrak{A}), 0, 0) \subset \mathfrak{A}, \quad N(T^*(\mathcal{H})) \subset \mathfrak{A}$$

and

$$\mathfrak{A} = (\pi(\mathfrak{A}), 0, 0) + N(T^*(\mathcal{H})).$$

(iv) *$\lambda(N(\mathfrak{A})) = \mathcal{H}$.*

Proof

(1) This is trivial.
(2) Suppose that $\mathfrak{A}$ is nondegenerate and singular. Then, since $N_{\mathfrak{A}} = I$, it follows that $\lambda(N(\mathfrak{A}))$ is dense in $\mathcal{H}$, which implies by the nondegenerateness of $\mathfrak{A}$ that $P_{\mathfrak{A}} = I$. Thus, $S_{\mathfrak{A}} = I$. The converse is trivial. Suppose, conversely, that $S_{\mathfrak{A}} = I$. Take an arbitrary $K \in \mathfrak{A}^{\tau}$. Since $P_{\mathfrak{A}^{\tau}} = I - N_{\mathfrak{A}} = 0$ by Theorem 4.1.7, we have $\pi(K) = 0$. Furthermore, since

$$\pi(A)\lambda(K) = \pi(K)\lambda(A) = 0 \quad \text{and} \quad \pi(A)\lambda(K^{\sharp}) = 0$$

for all $A \in \mathfrak{A}$, it follows from $P_{\mathfrak{A}} = I$ that $\lambda(K) = \lambda(K^{\sharp}) = 0$. Hence, $\mathfrak{A}^{\tau} = \{O\}$. Since $\mathfrak{A}^{\rho} \subset \mathfrak{A}^{\tau}$, we have $\mathfrak{A}^{\rho} = \{O\}$, which implies that $\mathfrak{A}^{\tau\tau} = \mathfrak{A}^{\rho\rho} = T^*(\mathcal{H})$. Therefore $\mathfrak{A}^{T} = \mathfrak{A}^{\pi\pi}$.
(3) The implications (iii)⇒(iv)⇒(i) are trivial.

(i)⇒(ii) Since $S_{\mathfrak{A}} = I$ by (2), it follows from Lemma 4.1.5, (1) that $(0, x, y^*) \in N(S_{\mathfrak{A}}) = P(\mathfrak{A}) \cap N(\mathfrak{A}) \subset \mathfrak{A}$ for all $x,\ y \in \mathcal{H}$, which proves (ii).
(ii)⇒(iii) Take an arbitrary $A \in \mathfrak{A}$. By assumption (ii) we have

$$(0, \lambda(A), \lambda(A^{\sharp})^*) \in \mathfrak{A}$$

and

$$(\pi(A), 0, 0) = A - (0, \lambda(A), \lambda(A^{\sharp})^*) \in \mathfrak{A}. \tag{4.2.18}$$

Thus, $(\pi(\mathfrak{A}), 0, 0) \subset \mathfrak{A}$. Take an arbitrary sequence $\{A_n\}$ in $\mathfrak{A}$ such that $\lim_{n\to\infty} \|\pi(A_n) - X_0\| = 0$ for some $X_0 \in B(\mathcal{H})$. Then, since

$$\lim_{n\to\infty} \|(\pi(A_n), 0, 0) - (X_0, 0, 0)\| = 0$$

and $\mathfrak{A}$ is a CT^*-algebra, it follows from (4.2.18) that $X := (X_0, 0, 0) \in \mathfrak{A}$ and $X_0 = \pi(X) \in \pi(\mathfrak{A})$. Thus $\pi(\mathfrak{A})$ is a C^*-algebra on $\mathcal{H}$, which proves (iii). This completes the proof.

We next characterize nilpotent CT^*-algebras.

Proposition 4.2.10 *Let $\mathfrak{A}$ be a CT^*-algebra on $\mathcal{H}$. Then the following statements are equivalent:*

(i) $\mathfrak{A}$ *is nilpotent.*
(ii) $\mathfrak{A} = N(\mathfrak{A})$.
(iii) $\mathfrak{A} = \nu(\mathfrak{A})$.
(iv) $\mathfrak{A} = Z(\mathfrak{A})$.

Proof The equivalence of (i), (ii) and (iii) is almost trivial.
(i)⇒(iv) By Theorem 4.1.4 and (iii) we have

$$Z(\mathfrak{A}) = \nu_{I-P_{\mathfrak{A}}}(\mathfrak{A}) = \nu(\mathfrak{A}) = \mathfrak{A}.$$

(iv)⇒(ii) Since

$$Z(\mathfrak{A}) = \nu_{I-P_{\mathfrak{A}}}(\mathfrak{A}) \subset N(\mathfrak{A}) \subset \mathfrak{A} = Z(\mathfrak{A}),$$

it follows that $N(\mathfrak{A}) = \mathfrak{A}$ and $P_{\mathfrak{A}} = 0$. Hence, $\mathfrak{A}$ is nilpotent.
This completes the proof.

We finally characterize nilpotent and singular CT^*-algebras. For that, we regard

$$\nu(\mathfrak{A}) = \{(0, \lambda(A), \lambda(A^\sharp)^*); \quad A \in \mathfrak{A}\}$$

as a subspace of the Hilbert space $\mathcal{H} \oplus \mathcal{H}^*$ with inner product:

$$\big((x_1, y_1^*)|(x_2, y_2^*)\big) = (x_1|x_2) + (y_2|y_1),$$

and denote by $\nu(\mathfrak{A})^\perp$ the orthogonal complement of $\nu(\mathfrak{A})$ in $\mathcal{H} \oplus \mathcal{H}^*$.

Lemma 4.2.11 *Suppose that $\mathfrak{A}$ is a nilpotent CT^*-algebra on $\mathcal{H}$. Then $\nu(\mathfrak{A})$ is a closed subspace in the Hilbert space $\mathcal{H} \oplus \mathcal{H}^*$ and $\nu(\mathfrak{A}^\rho) = J\nu(\mathfrak{A})^\perp$, where*

$$J = \begin{pmatrix} I & 0 \\ 0 & -I \end{pmatrix}.$$

Proof By Lemma 2.6.7 we have

$$\nu(\mathfrak{A}^\rho) \subset J\nu(\mathfrak{A})^\perp.$$

We show the converse. Take an arbitrary $(x, y^*) \in \nu(\mathfrak{A})^\perp$, and write $K = (0, x, -y^*)$. Then it is easily shown that $K \in \mathfrak{A}^\tau$ and $(\lambda(A)|x) = (-y|\lambda(A^\sharp))$ for all $A \in \mathfrak{A}$. Hence, $K \in \mathfrak{A}^\rho$ and $J(x, y^*) = \nu(K) \in \nu(\mathfrak{A}^\rho)$. Thus, $J\nu(\mathfrak{A})^\perp \subset \nu(\mathfrak{A}^\rho)$. This completes the proof.

Proposition 4.2.12 *Let $\mathfrak{A}$ be a CT^*-algebra on $\mathcal{H}$. Then the following statements are equivalent:*

(i) *$\mathfrak{A}$ is singular and nilpotent.*
(i)$'$ *$\mathfrak{A}^\tau$ is singular and nilpotent.*
(ii) *$\mathfrak{A}$ is nilpotent and $\lambda(\mathfrak{A})$ is dense in $\mathcal{H}$.*
(ii)$'$ *$\mathfrak{A}^\tau$ is nilpotent and $\lambda(\mathfrak{A}^\tau)$ is dense in $\mathcal{H}$.*

In this case, then $\mathfrak{A}^T = \mathfrak{A}$.

Proof The equivalence of (i) and (ii) is trivial. Furthermore, the equality:

$$Z_{\mathfrak{A}} = (I - P_{\mathfrak{A}})N_{\mathfrak{A}} = N_{\mathfrak{A}^\tau}(I - P_{\mathfrak{A}^\tau}) = Z_{\mathfrak{A}^\tau}$$

implies that statements (i), (ii), (i)$'$ and (ii)$'$ are equivalent. Suppose that $\mathfrak{A}$ is singular and nilpotent. Then $\mathfrak{A} = N(\mathfrak{A})$ and $\lambda(\mathfrak{A})$ is dense in $\mathcal{H}$ by Proposition 4.2.9, so that

$$\mathfrak{A}^\pi = T^*(\mathcal{H}) \quad \text{and} \quad \mathfrak{A}^{\pi\pi} = N(T^*(\mathcal{H})) \tag{4.2.19}$$

and

$$\mathfrak{A}^\tau = N(T^*(\mathcal{H})). \tag{4.2.20}$$

Therefore

$$\mathfrak{A}^\rho = \nu(\mathfrak{A}^\rho) = N(\mathfrak{A}^\rho). \tag{4.2.21}$$

Furthermore, it follows from Lemma 4.2.10 that

$$\mathfrak{A}^\rho = \nu(\mathfrak{A}^\rho) = J\nu(\mathfrak{A})^\perp$$

and

$$\begin{aligned} \nu(\mathfrak{A}^{\rho\rho}) &= J\nu(\mathfrak{A}^\rho)^\perp = J(J\nu(\mathfrak{A})^\perp)^\perp \\ &= \nu(\mathfrak{A})^{\perp\perp} = \nu(\mathfrak{A}) \\ &= \mathfrak{A}. \end{aligned} \tag{4.2.22}$$

Since $\mathfrak{A}^\tau$ is singular and nilpotent, it follows, as in (4.2.20), that

$$\mathfrak{A}^{\tau\tau} = N(T^*(\mathcal{H})). \tag{4.2.23}$$

By (4.2.19), (4.2.22) and (4.2.23) we have

$$\mathfrak{A}^T = N(T^*(\mathcal{H})) \cap \mathfrak{A}^{\rho\rho} = \nu(\mathfrak{A}^{\rho\rho}) = \mathfrak{A}.$$

This completes the proof.

In case that a CT^*-algebra $\mathfrak{A}$ on $\mathcal{H}$ is isomorphic to a CT^*-algebra $\mathfrak{B}$ on $\mathcal{K}$, it is natural to consider whether the regularity (resp. the semisimplicity, the nondegenerateness, the singularity, the nilpotentness) of $\mathfrak{A}$ and $\mathfrak{B}$ is equivalent. For this question we get the following

Proposition 4.2.13 *Let $\mathfrak{A}$ be a CT^*-algebra on $\mathcal{H}$ and $\mathfrak{B}$ a CT^*-algebra on $\mathcal{K}$. Suppose that $\mathfrak{A}$ is isomorphic to $\mathfrak{B}$. Then $\mathfrak{A}$ is semisimple (resp. nondegenerate, singular, nilpotent) if and only if $\mathfrak{B}$ is semisimple (resp. nondegenerate, singular, nilpotent). However, the regularity of $\mathfrak{A}$ and $\mathfrak{B}$ is not necessarily equivalent.*

Proof Let Φ be an isometric $*$-isomorphism and isometry of $\mathfrak{A}[\tau_u]$ onto $\mathfrak{B}[\tau_u]$. Since

$$\|\pi(\Phi(A))\| = \|\pi(A)\| \quad \text{and} \quad \|\lambda(\Phi(A))\| = \|\lambda(A)\|$$

for all $A \in \mathfrak{A}$, it follows that $P(\mathfrak{B}) = \Phi(P(\mathfrak{A}))$ and $N(\mathfrak{B}) = \Phi(N(\mathfrak{A}))$, which implies that the semisimplicity (resp. nondegenerateness, singularity, nilpotentness) of $\mathfrak{A}$ and $\mathfrak{B}$ is equivalent. In the next Example 4.2.14 we give a semisimple and nonregular CT^*-algebra which is isomorphic to a regular CT^*-algebra. This completes the proof.

Example 4.2.14 Let $\mathfrak{A}$ be a CT^*-algebra on $\mathcal{H}$ such that $0 \lneqq P_{\mathfrak{A}} \lneqq I$. Then $\tau_{P_{\mathfrak{A}}}(\mathfrak{A})$ is a CT^*-algebra on $\mathcal{H}$ which is not regular. We define a T^*-algebra $\mathfrak{A}_{P_{\mathfrak{A}}}$ on $P_{\mathfrak{A}}\mathcal{H}$ by

$$\mathfrak{A}_{P_{\mathfrak{A}}} = \left\{ A_{P_{\mathfrak{A}}} := \left(\pi(A)\lceil_{P_{\mathfrak{A}}\mathcal{H}}, P_{\mathfrak{A}}\lambda(A), (P_{\mathfrak{A}}\lambda(A^\sharp))^*\right); \quad A \in \mathfrak{A} \right\}.$$

Then it is easily shown that the map Φ defined by

$$\Phi : \quad A \in \tau_{P_{\mathfrak{A}}}(\mathfrak{A}) \longrightarrow A_{P_{\mathfrak{A}}} \in \mathfrak{A}_{P_{\mathfrak{A}}}$$

is a $*$-isomorphism of the CT^*-algebra $\tau_{P_{\mathfrak{A}}}(\mathfrak{A})$ onto the T^*-algebra $\mathfrak{A}_{P_{\mathfrak{A}}}$. Furthermore, since

$$\begin{aligned} \|\pi(A)\| &= \sup\{\|\pi(A)x\|; \quad \|x\| \leqq 1\} \\ &= \sup\{\|\pi(A)P_{\mathfrak{A}}x\|; \quad \|x\| \leqq 1\} \end{aligned}$$

$$\leqq \sup\{\|\pi(A)P_{\mathfrak{A}}x\|;\quad \|P_{\mathfrak{A}}x\| \leqq 1\}$$
$$= \|\pi(A)\lceil_{P_{\mathfrak{A}}\mathcal{H}}\|$$
$$\leqq \|\pi(A)\|$$

for all $A \in \mathfrak{A}$, it follows that Φ is an isometry. Hence, $\mathfrak{A}_{P_{\mathfrak{A}}}$ is a regular CT^*-algebra on $P_{\mathfrak{A}}\mathcal{H}$ which is isomorphic to the nonregular CT^*-algebra $\tau_{P_{\mathfrak{A}}}(\mathfrak{A})$ on $\mathcal{H}$.

4.3 Commutative Semisimple CT^*-Algebras

In this section we shall show that every commutative semisimple CT^*-algebra $\mathfrak{A}[\tau_u]$ is isomorphic to a regular CT^*-algebra $\left(C_0(\Omega) \cap L^2(\Omega, \mu)\right)[\tau_u]$. Here $C_0(\Omega)$ is the commutative C^*-algebra of continuous functions f on a locally compact space Ω vanishing at infinity on Ω equipped with the usual function operations: $f+g$, αf, fg, the involution f^* ($f^*(\lambda) = \overline{f(\lambda)}$) and the C^*-norm $\|f\|_u := \sup_{x\in\Omega}|f(x)|$, and $L^2(\Omega, \mu)$ is the L^2-space defined by a regular Borel measure μ on Ω. Then $C_0(\Omega) \cap L^2(\Omega, \mu)$ is a Banach $*$-algebra with the above function operations $f+g$, αf, fg, the involution f^* and the norm $\|f\| := \max(\|f\|_u, \|f\|_2)$, and it is regarded as the regular commutative CT^*-algebra:

$$\tau\left(C_0(\Omega) \cap L^2(\Omega, \mu)\right) := \{\tau(f) = (\pi(f), \lambda(f), \lambda(f^*)^*);\quad f \in C_0(\Omega) \cap L^2(\Omega, \mu)\},$$

where $\pi(f)$ is a bounded operator on $L^2(\Omega, \mu)$ defined by $\pi(f)g = fg$, $g \in L^2(\Omega, \mu)$ and $\lambda(f) = f$. Then we get the following main theorem of this section.

Theorem 4.3.1 *Let $\mathfrak{A}$ be a commutative semisimple CT^*-algebra on $\mathcal{H}$. Then there exist a locally compact space Ω and a regular Borel measure μ on Ω such that $\mathfrak{A}[\tau]$ is isomorphic to the commutative regular CT^*-algebra $\left(C_0(\Omega) \cap L^2(\Omega, \mu)\right)[\tau_u]$.*

Since $\overline{\pi(\mathfrak{A})}[\tau_u^\pi]$ of the uniform closure of $\pi(\mathfrak{A})$ is a commutative C^*-algebra, it follows from ([37, Theorem 4.4]) that there exists a locally compact space Ω such that it is isomorphic to the commutative C^*-algebra $C_0(\Omega)$. Denote the isometric $*$-isomorphism of $\overline{\pi(\mathfrak{A})}[\tau_u^\pi]$ onto $C_0(\Omega)$ by φ. Let $C_c(\Omega)$ be the subset of continuous functions with compact support on Ω. The support of a function f is denoted by supp f. We will prove Theorem 4.3.1 in the following process.

Step 1 $C_c(\Omega) \subset \varphi(\pi(\mathfrak{A}))$.

Proof For any $f \in C_0(\Omega)$ there exists a sequence $\{A_n\}$ in $\mathfrak{A}$ such that $\lim_{n\to\infty} \|\pi(A_n) - \varphi^{-1}(f)\| = 0$. Since $\mathfrak{A}$ is commutative, we have

$$\pi(B)^*\lambda(A_n^\sharp) = \lambda(B^\sharp A_n^\sharp) = \lambda(A_n^\sharp B^\sharp) = \pi(A_n^*)\lambda(B^\sharp)$$

for all $B \in \mathfrak{A}$. Hence it follows that

$$\begin{aligned} A_n B &= \left(\pi(A_n)\pi(B), \pi(A_n)\lambda(B), (\pi(B)^*\lambda(A_n^\sharp))^*\right) \\ &= \left(\pi(A_n)\pi(B), \pi(A_n)\lambda(B), (\pi(A_n)^*\lambda(B^\sharp))^*\right), \end{aligned}$$

and

$$\tau_u\text{-}\lim_{n\to\infty} A_n B = \left(\varphi^{-1}(f)\pi(B), \varphi^{-1}(f)\lambda(B), (\varphi^{-1}(f)^*\lambda(B^\sharp))^*\right).$$

The right hand side can be interpreted as an action of $\varphi^{-1}(C_0(\Omega))$ to $\mathfrak{A}$, so that it will be denoted by $\varphi^{-1}(f)B$. Therefore,

$$\varphi^{-1}(f)\mathfrak{A} \subset P(\mathfrak{A}) \tag{4.3.1}$$

for all $f \in C_0(\Omega)$. For any $x \in \Omega$ there exists an element A of $\mathfrak{A}$ such that $\varphi(\pi(A))(x) \neq 0$, so $\varphi(\pi(A^\sharp A)(x) = |\varphi(\pi(A))x|^2 > 0$. Hence, for any $f \in C_c(\Omega)$ there exists an element B of $\mathfrak{A}$ such that

$$\begin{aligned} &\varphi(\pi(B))(x) > 0 \quad \text{on} \quad x \in \operatorname{supp} f, \\ &\varphi(\pi(B))(x) \geqq 0 \quad \text{on} \quad x \in \Omega \setminus \operatorname{supp} f, \end{aligned}$$

so that there exists an element h of $C_c(\Omega)$ such that $h(x) = (\varphi(\pi(B))(x))^{-1}$ for all $x \in \operatorname{supp} f$. Let $A = \varphi^{-1}(g)B$. Then $A \in \mathfrak{A}$ by (4.3.1) and

$$\begin{aligned} f &= fh\varphi(\pi(B)) \\ &= \varphi(\varphi^{-1}(fh)\pi(B)) \\ &= \varphi(\pi(A)) \in \varphi(\pi(\mathfrak{A})). \end{aligned}$$

In what follows we denote the set $\{A \in \mathfrak{A};\ \varphi(\pi(A)) \in C_c(\Omega)\}$ by $\mathfrak{A}_c$. Then the mapping

$$\Phi: \quad A \in \mathfrak{A}_c \longrightarrow \varphi(\pi(A)) \in C_c(\Omega)$$

is surjective by Step 1, and is injective by the semisimplicity of $\mathfrak{A}$. Furthermore, we can easily show the following

Step 2 $\mathfrak{A}_c$ is a $*$-subalgebra of the CT^*-algebra $\mathfrak{A}$, $C_c(\Omega)$ is a $*$-subalgebra of the CT^*-algebra $C_0(\Omega) \cap L^2(\Omega, \mu)$ and Φ is an algebraic $*$-isomoprphism of the T^*-algebra $\mathfrak{A}_c$ onto the T^*-algebra $C_c(\Omega)$.

We will verify that Φ is an isometry of the T^*-algebra $\mathfrak{A}_c[\tau_u]$ on the T^*-algebra $C_c(\Omega)[\tau_u]$. Since

$$\begin{aligned} \|\Phi(A)\| &= \max\left(\|\varphi(\pi(A))\|_u, \|\varphi(\pi(A))\|_2\right), \\ \|A\| &= \max\left(\|\pi(A)\|, \|\lambda(A)\|, \|\lambda(A^\sharp)\|\right) \end{aligned}$$

and $\|\varphi(\pi(A))\|_u = \|\pi(A)\|$ for all $A \in \mathfrak{A}_c$, it suffices to show that

$$\|\varphi(\pi(A))\|_2 = \|\lambda(A)\|.$$

We will prove this equality after some preparations.

Step 3 For any $A \in \mathfrak{A}_c$ there exists an element B of $\mathfrak{A}_c$ such that

$$BA = A \quad \text{and} \quad B^\sharp A = A.$$

Proof For any $A \in \mathfrak{A}_c$ there exists an element B of $\mathfrak{A}_c$ such that $\varphi(\pi(B))(x) = 1$ for all $x \in \operatorname{supp} \varphi(\pi(A))$. Then

$$\varphi(\pi(BA)) = \varphi(\pi(B))\varphi(\pi(A)) = \varphi(\pi(A));$$

hence $\pi(BA) = \pi(A)$. Therefore, for any $C \in \mathfrak{A}^\tau$ and $x \in \mathcal{H}$, we see that

$$\begin{aligned} (\pi(B)\lambda(A)|\pi(C)x) &= (\pi(BA)\lambda(C^\sharp)|x) \\ &= (\pi(A)\lambda(C^\sharp)|x) \\ &= (\lambda(A)|\pi(C)x), \end{aligned}$$

which implies since $\pi(\mathfrak{A}^\tau)$ is nondegenerate that $\pi(B)\lambda(A) = \lambda(A)$. Similarly, $\pi(B)^*\lambda(A^\sharp) = \lambda(A^\sharp)$. Hence we have

$$\begin{aligned} BA &= (\pi(BA), \pi(B)\lambda(A), (\pi(B)^*\lambda(A^\sharp))^*) \\ &= (\pi(A), \lambda(A), \lambda(A^\sharp)^*) \\ &= A. \end{aligned}$$

Since $\varphi(\pi(B^\sharp))(x) = \overline{\varphi(\pi(B))(x)}$, it follows that $\varphi(\pi(B))(x) = 1$ for $x \in \operatorname{supp} \varphi(\pi(A))$. Hence we can show $B^\sharp A = A$ in the same way as above.

Step 4 We define a function F on $C_c(\Omega)$ by

$$F(f) = (\lambda(\varphi^{-1}(f)B)|\lambda(B)), \quad f \in C_c(\Omega).$$

Then it is determined independently from the choice of $B \in \mathfrak{A}_c$ which satisfies $\varphi(\pi(B))(x) = 1$ for all $x \in \text{supp } f$. Moreover, it is a positive linear functional on $C_c(\Omega)$.

Proof Let $f \in C_c(\Omega)$. By Step 1 $f = \varphi(\pi(A))$ for some $A \in \mathfrak{A}_c$. Take arbitrary $B, B_1 \in \mathfrak{A}_c$ satisfying $\varphi(\pi(B))(x) = \varphi(\pi(B_1))(x) = 1$ for all $x \in \text{supp } f$, and take an element $h \in C_c(\Omega)$ which $h(x) = 1$ for all $x \in \text{supp } \varphi(\pi(B)) \cup \text{supp } \varphi(\pi(B_1))$. Then, $h = \varphi(\pi(C))$ for some $C \in \mathfrak{A}_c$, so that it follows from Step 3 that

$$BA = B^{\sharp}A = A, \quad B_1A = B_1^{\sharp}A = A, \quad BC = B \quad \text{and} \quad B_1C = B_1. \tag{4.3.2}$$

Hence we have

$$\begin{aligned}
(\lambda(\varphi^{-1}(f)B)|\lambda(B)) &= (\lambda(AB)|\lambda(B)) \\
&= (\lambda(A)|\lambda(BC)) \\
&= (\lambda(B^{\sharp}A)|\lambda(C)) \\
&= (\lambda(A)|\lambda(C)).
\end{aligned}$$

Similarly, $(\lambda(\varphi^{-1}(f)B_1)|\lambda(B_1)) = (\lambda(A)|\lambda(C))$. Thus, the value of $F(f)$ does not depend on the choice of B. We next show that F is a positive linear functional on $C_c(\Omega)$. Take arbitrary $f, g \in C_c(\Omega)$ and take an element h of $C_c(\Omega)$ with $h = 1$ on $\text{supp } f \cup \text{supp } g \cup \text{supp } (f+g)$. Since $f = \varphi(\pi(A))$, $g = \varphi(\pi(B))$ and $h = \varphi(\pi(C))$ for some $A, B, C \in \mathfrak{A}_c$, we see that

$$\begin{aligned}
F(f+g) &= (\varphi^{-1}(f+g)\lambda(C)|\lambda(C)) \\
&= (\varphi^{-1}(f)\lambda(C)|\lambda(C)) + (\varphi^{-1}(g)\lambda(C)|\lambda(C)) \\
&= F(f) + F(g), \\
F(\alpha f) &= (\varphi^{-1}(\alpha f)\lambda(C)|\lambda(C)) \\
&= \alpha(\varphi^{-1}(f)\lambda(C)|\lambda(C)) \\
&= \alpha F(f), \\
F(f^* f) &= (\varphi^{-1}(f^* f)\lambda(C)|\lambda(C)) \\
&= (\varphi^{-1}(f^*)\varphi^{-1}(f)\lambda(C)|\lambda(C)) \\
&= \|\varphi^{-1}(f)\lambda(C)\|^2 \geqq 0
\end{aligned}$$

fow all $f, g \in C_c(\Omega)$ and $\alpha \in \mathbb{C}$. Thus F is a positive linear functional on $C_c(\Omega)$.

Step 5 $\|\varphi(\pi(A))\|_2 = \|\lambda(A)\|$ for all $A \in \mathfrak{A}_c$.

Proof For any $A, \ B \in \mathfrak{A}_c$ choose an element C of $\mathfrak{A}_c$ with $\varphi(\pi(C)) = 1$ on supp $\varphi(\pi(A)) \cup$ supp $\varphi(\pi(B)) \cup$ supp $\varphi(\pi(B^\sharp A))$. Then it follows from (4.3.2) and the Riesz-Markov theorem [29] that

$$\begin{aligned}(\lambda(A)|\lambda(B)) &= (\lambda(AC)|\lambda(BC)) \\ &= (\pi(B^\sharp A)\lambda(C)|\lambda(C)) \\ &= F(\varphi(\pi(B^\sharp A))) \\ &= \int_\Omega \varphi(\pi(B^\sharp A))(\lambda) d\mu(\lambda) \\ &= \int_\Omega \overline{\varphi(\pi(B))(\lambda)} \varphi(\pi(A))(\lambda) d\mu(\lambda) \\ &= (\varphi(\pi(A))|\varphi(\pi(B))),\end{aligned}$$

which completes the proof.
Thus, Φ is an isometric $*$-isomorphism of the T^*-algebra $\mathfrak{A}_c[\tau_u]$ onto the T^*-algebra $C_c(\Omega)[\tau_u]$.

Proof of Theorem 4.3.1 We show that Φ can be extended to the $*$-isomorphism of the CT^*-algebra $\mathfrak{A}[\tau_u]$ onto the CT^*-algebra $\left(C_0(\Omega) \cap L^2(\Omega, \mu)\right)[\tau_u]$. By Step 2, 4 it suffices to show that $\bar{\mathfrak{A}}_c[\tau_u] = \mathfrak{A}$ and $\overline{C_c(\Omega)}[\tau_u] = C_0(\Omega) \cap L^2(\Omega, \mu)$. Indeed, take arbitrary $A, \ B \in \mathfrak{A}$. Then, $\pi(A) = \varphi^{-1}(f)$ and $\pi(B) = \varphi^{-1}(g)$ for some $f, \ g \in C_0(\Omega)$. Now, we can take a sequence $\{f_n\}$ in $C_c(\Omega)$ with $\lim_{n\to\infty} \|f_n - f\|_u = 0$, and then there exists a sequence $\{A_n\}$ in $\mathfrak{A}_c$ with $\pi(A_n) = \varphi^{-1}(f_n)$ and

$$\lim_{n\to\infty} \|\pi(A_n) - \pi(A)\| = \lim_{n\to\infty} \|f_n - f\|_u = 0.$$

Since $\{f_n g\}$ is a sequence in $C_c(\Omega)$, it follows that

$$\{A_n B\} \subset \mathfrak{A}_c \quad \text{and} \quad \lim_{n\to\infty} \|A_n B - AB\| = 0,$$

which implies that $\mathfrak{A}_c$ is τ_u-dense in $P(\mathfrak{A})$. By the semisimplicity of $\mathfrak{A}$, we have $\mathfrak{A} = P(\mathfrak{A})$; hence $\bar{\mathfrak{A}}_c[\tau_u] = \mathfrak{A}$. On the other hand it is clear that $\overline{C_c(\Omega)}[\tau_u] = C_0(\Omega) \cap L^2(\Omega, \mu)$. Thus the CT^*-algebra $\mathfrak{A}[\tau_u]$ is isomorphic to the CT^*-algebra $(C_0(\Omega) \cap L^2(\Omega, \mu))[\tau_u]$. This completes the proof.

The next result is now immediate from Theorem 4.3.1.

Corollary 4.3.2 *Suppose that $\mathfrak{A}$ is a commutative CT^*-algebra on $\mathcal{H}$. Then there exists a locally compact space Ω and a regular Borel measure μ on Ω such that the CT^*-algebra $\tau_{F_\mathfrak{A}}(\mathfrak{A})$ is isomorphic to the CT^*-algebra $(C_0(\Omega) \cap L^2(\Omega, \mu))$.*

4.4 Some Results Obtained from T^*-Algebras

4.4.1 Projections Defined by T^*-Algebras

Let $\mathfrak{A}$ be a T^*-algebra on $\mathcal{H}$ and denote the uniform closure $\tilde{\mathfrak{A}}[\tau_u]$ of $\mathfrak{A}$ by $\tilde{\mathfrak{A}}$ for simplicity. Then $\tilde{\mathfrak{A}}$ is a CT^*-algebra on $\mathcal{H}$ satisfying

$$P(\mathfrak{A}) = P(\tilde{\mathfrak{A}}), \quad N(\mathfrak{A}) \subset N(\tilde{\mathfrak{A}}) \quad \text{and} \quad Z(\mathfrak{A}) \subset Z(\tilde{\mathfrak{A}}).$$

Hence, for the relationship between the projections defined by $\mathfrak{A}$ and $\tilde{\mathfrak{A}}$ we have

$$\begin{aligned} P_{\mathfrak{A}} = P_{\tilde{\mathfrak{A}}}, \quad N_{\mathfrak{A}} \leqq N_{\tilde{\mathfrak{A}}}, \quad F_{\mathfrak{A}} &:= P_{\mathfrak{A}}(I - N_{\mathfrak{A}}) \geqq F_{\tilde{\mathfrak{A}}}, \\ S_{\mathfrak{A}} := P_{\mathfrak{A}} N_{\mathfrak{A}} \leqq S_{\tilde{\mathfrak{A}}}, \quad Z_{\mathfrak{A}} &:= (I - P_{\mathfrak{A}}) N_{\mathfrak{A}} \leqq Z_{\tilde{\mathfrak{A}}}, \\ V_{\mathfrak{A}} &:= (I - P_{\mathfrak{A}})(I - N_{\mathfrak{A}}) \geqq V_{\tilde{\mathfrak{A}}}. \end{aligned} \tag{4.4.1}$$

Furthermore, since $\mathfrak{A}^\tau = \tilde{\mathfrak{A}}^\tau$, it follows from Theorem 4.1.7 and (4.4.1) that

$$\begin{aligned} P_{\mathfrak{A}^\tau} = P_{\tilde{\mathfrak{A}}^\tau} = I - N_{\tilde{\mathfrak{A}}} \leqq I - N_{\mathfrak{A}}, \quad & N_{\mathfrak{A}^\tau} = N_{\tilde{\mathfrak{A}}^\tau} = I - P_{\tilde{\mathfrak{A}}} = I - P_{\mathfrak{A}}, \\ F_{\mathfrak{A}^\tau} = F_{\tilde{\mathfrak{A}}^\tau} = F_{\tilde{\mathfrak{A}}} \leqq F_{\mathfrak{A}}, \quad & S_{\mathfrak{A}^\tau} = S_{\tilde{\mathfrak{A}}^\tau} = V_{\tilde{\mathfrak{A}}} \leqq V_{\mathfrak{A}}, \\ Z_{\mathfrak{A}^\tau} = Z_{\tilde{\mathfrak{A}}^\tau} = Z_{\tilde{\mathfrak{A}}} \geqq Z_{\mathfrak{A}}, \quad & V_{\mathfrak{A}^\tau} = V_{\tilde{\mathfrak{A}}^\tau} = S_{\tilde{\mathfrak{A}}} \geqq S_{\mathfrak{A}}. \end{aligned} \tag{4.4.2}$$

We now define the notions of semisimplicity, regularity and singularity of T^*-algebras.

Definition 4.4.1 A T^*-algebra $\mathfrak{A}$ is called nondegenerate (resp. semisimple, regular, singular, nilpotent) if $P_{\mathfrak{A}} = I$ (resp. $P_{\mathfrak{A}^\tau} = I$, $P_{\mathfrak{A}} = P_{\mathfrak{A}^\tau} = I$, $P_{\mathfrak{A}^\tau} = 0$, $P_{\mathfrak{A}} = 0$).

From (4.4.1) and (4.4.2) it follows that $\mathfrak{A}$ is nondegenerate (resp. semisimple, regular, singular) if and only if $\tilde{\mathfrak{A}}$ is nondegenerate (resp. semisimple, regular, singular).

4.4.2 The Vector Representation of the CT^*-Algebra Generated by a T^*-Algebra

In this subsection we investigate the relationship between the vector representation $\lambda(\mathfrak{A})$ of a T^*-algebra $\mathfrak{A}$ and that of the CT^*-algebra $\tilde{\mathfrak{A}}$.

Theorem 4.4.2 *Let $\mathfrak{A}$ be a T^*-algebra on $\mathcal{H}$. For $A \in \tilde{\mathfrak{A}}$ we put*

$$\mathfrak{F}(A) = \left\{ \{A_n\} \subset \mathfrak{A}; \quad \sum_{n=1}^{\infty} \|\pi(A_n)\|^2 < \infty \right.$$
$$\left. \text{and} \quad \sum_{n=1}^{\infty} \pi(A_n)^* \pi(A_n) \geqq \pi(A)^* \pi(A) \right\}.$$

Then, $\mathfrak{F}(A) \neq \emptyset$ and

$$\|F_{\tilde{\mathfrak{A}}} \lambda(A)\|^2 = \inf \left\{ \sum_{n=1}^{\infty} \|F_{\tilde{\mathfrak{A}}} \lambda(A_n)\|^2; \quad \{A_n\} \in \mathfrak{F}(A) \right\}.$$

In particular, if $\mathfrak{A}$ is semisimple, then

$$\|\lambda(A)\|^2 = \inf \left\{ \sum_{n=1}^{\infty} \|\lambda(A_n)\|^2; \quad \{A_n\} \in \mathfrak{F}(A) \right\}.$$

We prove this theorem in the following processes.

Step 1 For any $A \in \tilde{\mathfrak{A}}$ and $\{A_n\} \in \mathfrak{F}(A)$ we have

$$\sum_{n=1}^{\infty} \|F_{\tilde{\mathfrak{A}}} \lambda(A_n)\|^2 \geqq \|F_{\tilde{\mathfrak{A}}} \lambda(A)\|^2.$$

Proof Let $A \in \tilde{\mathfrak{A}}$ and $\{A_n\} \in \mathfrak{F}(A)$. Then it follows from Corollary 4.2.7 that

$$\begin{aligned} \sum_{n=1}^{\infty} \|F_{\tilde{\mathfrak{A}}} \lambda(A_n)\|^2 &\geqq \sum_{n=1}^{\infty} \|\pi(A_n)\lambda(K)\|^2 \\ &\geqq \|\pi(A)\lambda(K)\|^2 \end{aligned}$$

for all $K \in u_\pi \left(\tau_{F_{\tilde{\mathfrak{A}}}}(\tilde{\mathfrak{A}})^\tau \right)$, which implies by Corollary 4.2.7 again that

$$\sum_{n=1}^{\infty} \|F_{\tilde{\mathfrak{A}}} \lambda(A_n)\|^2 \geqq \|F_{\tilde{\mathfrak{A}}} \lambda(A)\|^2.$$

From Step 1 the proof of Theorem 4.4.2 completes if for any $A \in \tilde{\mathfrak{A}}$ and $\varepsilon > 0$, we can find an element $\{A_n\}$ of $\mathfrak{F}(A)$ such that

$$\|F_{\tilde{\mathfrak{A}}}\lambda(A)\|^2 + \varepsilon > \sum_{n=1}^{\infty} \|F_{\tilde{\mathfrak{A}}}\lambda(A_n)\|^2.$$

In the following we use the following inequality several times. Let a and b be elements of a $*$-algebra and δ a positive scalar. Since $a = b + (a - b)$, it follows that

$$\begin{aligned} a^*a &\leqq (b + (a-b))^* (b + (a-b)) + \left(\delta^{-\frac{1}{2}} b - \delta^{-\frac{1}{2}}(a-b)\right)^* \left(\delta^{\frac{1}{2}} b - \delta^{-\frac{1}{2}}(a-b)\right) \\ &= (1+\delta)b^*b + (1+\delta^{-1})(a-b)^*(a-b). \end{aligned} \tag{4.4.3}$$

In particular, if $\delta = 1$, then

$$a^*a \leqq 2(b^*b + (a-b)^*(a-b)). \tag{4.4.4}$$

Take arbitrary $A \in \tilde{\mathfrak{A}}$ and $\varepsilon > 0$.

Step 2 There exist a first approximation $A_1 \in \mathfrak{A}$ of A and the rest $B_1 \in \tau_{F_{\tilde{\mathfrak{A}}}}(\tilde{\mathfrak{A}})$ satisfying

(1) $\pi(A)^*\pi(A) \leqq \pi(A_1)^*\pi(A_1) + \pi(B_1)^*\pi(B_1)$,
(2) $\|\pi(A_1)\|^2 \leqq \|\pi(A)\|^2 + \frac{\varepsilon}{2}$ and $\|\pi(B_1)\|^2 < \frac{\varepsilon}{2^5}$,
(3) $\|F_{\tilde{\mathfrak{A}}}\lambda(A_1)\|^2 \leqq \|F_{\tilde{\mathfrak{A}}}\lambda(A)\|^2 + \frac{\varepsilon}{2}$ and $\|\lambda(B_1)\|^2 < \frac{\varepsilon}{2^5}$.

Proof

(1) For any $\varepsilon_1 > 0$ we choose $X_1 \in \mathfrak{A}$ with $\|\tau_{F_{\tilde{\mathfrak{A}}}}(A - X_1)\| < \varepsilon_1$, namely

$$\|\pi(A) - \pi(X_1)\| < \varepsilon_1, \tag{4.4.5}$$

$$\|F_{\tilde{\mathfrak{A}}}\lambda(A) - F_{\tilde{\mathfrak{A}}}\lambda(X_1)\| < \varepsilon_1. \tag{4.4.6}$$

We here put

$$A_s = \tau_{F_{\tilde{\mathfrak{A}}}}(A) \quad \text{and} \quad (X_1)_s = \tau_{F_{\tilde{\mathfrak{A}}}}(X_1),$$

and for any $\delta > 0$,

$$A_1 = (1+\delta)^{\frac{1}{2}} X_1 \quad \text{and} \quad B_1 = (1+\delta^{-1})^{\frac{1}{2}}(A_s - (X_1)_s).$$

Then, $A_1 \in \mathfrak{A}$, $B_1 \in \tau_{F_{\tilde{\mathfrak{A}}}}(\tilde{\mathfrak{A}})$. Since $A_s = (X_1)_s + (A_s - (X_1)_s)$, $\pi(A_s) = \pi(A)$ and $\pi((X_1)_s) = \pi(X_1)$, it follows from (4.4.3) that

$$\begin{aligned}\pi(A)^*\pi(A) &\leqq (1+\delta)\pi(X_1)^*\pi(X_1) + (1+\delta^{-1})\pi(A_s - (X_1)_s)^*\pi(A_s - (X_1)_s)\\ &= \pi(A_1)^*\pi(A_1) + \pi(B_1)^*\pi(B_1).\end{aligned}$$

(2) Since $X_1 = A + (X_1 - A)$, we have

$$\pi(X_1)^*\pi(X_1) \leqq (1+\delta)\pi(A)^*\pi(A) + (1+\delta^{-1})\pi(X_1 - A)^*\pi(X_1 - A).$$

Since $A_1 = (1+\delta)^{\frac{1}{2}} X_1$, it follows from (4.4.5) that

$$\begin{aligned}\pi(A_1)^*\pi(A_1) &= (1+\delta)\pi(X_1)^*\pi(X_1)\\ &\leqq (1+\delta)^2\pi(A)^*\pi(A) + (1+\delta)(1+\delta^{-1})\pi(X_1 - A)^*\pi(X_1 - A)\\ &< \pi(A)^*\pi(A) + \left\{(2\delta+\delta^2)\pi(A)^*\pi(A) + (1+\delta)(1+\delta^{-1})\varepsilon_1^2 I\right\},\end{aligned}$$

and that

$$\begin{aligned}\|\pi(B_1)\|^2 &= (1+\delta^{-1})\|\pi(A) - \pi(X_1)\|^2\\ &\leqq (1+\delta^{-1})\varepsilon_1^2.\end{aligned}$$

We now choose ε_1 and δ as follows:

$$(2\delta+\delta^2)\max\left(\|\pi(A)\|^2, \|F_{\tilde{\mathfrak{A}}}\lambda(A)\|^2\right) + (1+\delta)(1+\delta^{-1})\varepsilon_1^2 < \frac{\varepsilon}{2},$$

$$(1+\delta^{-1})\varepsilon_1^2 < \frac{\varepsilon}{2^5}. \tag{4.4.7}$$

Then we have

$$\|\pi(A_1)\|^2 < \|\pi(A)\|^2 + \frac{\varepsilon}{2} \text{ and } \|\pi(B_1)\|^2 < \frac{\varepsilon}{2^5}.$$

(3) Since

$$\begin{aligned}F_{\tilde{\mathfrak{A}}}\pi(K)\lambda(A) &= P_{\mathfrak{A}}\pi(K)\lambda(A)\\ &= P_{\mathfrak{A}}\pi(A)\lambda(K)\\ &= \pi(A)\lambda(K)\end{aligned}$$

for all $A \in \mathfrak{A}$ and $K \in \mathfrak{A}^\tau$, it follows from (4.4.9) that

$$\begin{aligned}\|F_{\tilde{\mathfrak{A}}}\pi(e'_\gamma)\lambda(A_1)\|^2 &= \|\pi(A_1)\lambda(e'_\gamma)\|^2 \\ &\leqq (1+\delta)^2\|\pi(A)\lambda(e'_\gamma)\|^2 + (1+\delta)(1+\delta^{-1})\|\pi(X_1 - A)\lambda(e'_\gamma)\|^2 \\ &= (1+\delta)^2\|F_{\tilde{\mathfrak{A}}}\pi(A)\lambda(e'_\gamma)\|^2 + (1+\delta)(1+\delta^{-1})\|F_{\tilde{\mathfrak{A}}}\pi(X_1 - A)\lambda(e'_\gamma)\|^2 \\ &= (1+\delta)^2\|F_{\tilde{\mathfrak{A}}}\pi(e'_\gamma)\lambda(A)\|^2 + (1+\delta)(1+\delta^{-1})\|F_{\tilde{\mathfrak{A}}}\pi(e'_\gamma)\lambda(X_1 - A)\|^2,\end{aligned}$$

where $\{e'_\gamma\}$ is a net in $u_\pi(\mathfrak{A}^\tau)$ such that $\pi(e'_\gamma)$ converges strongly to $P_{\mathfrak{A}^\tau}$, which yields by (4.4.6) and (4.4.7) that

$$\begin{aligned}\|F_{\tilde{\mathfrak{A}}}\lambda(A_1)\|^2 &\leqq (1+\delta)^2\|F_{\tilde{\mathfrak{A}}}\lambda(A)\|^2 + (1+\delta)(1+\delta^{-1})\|F_{\tilde{\mathfrak{A}}}\lambda(X_1 - A)\|^2, \\ &< \|F_{\tilde{\mathfrak{A}}}\lambda(A)\|^2 + ((2\delta+\delta^2)\|F_{\tilde{\mathfrak{A}}}\lambda(A)\|^2 + (1+\delta)(1+\delta^{-1})\varepsilon_1^2) \\ &< \|F_{\tilde{\mathfrak{A}}}\lambda(A)\|^2 + \frac{\varepsilon}{2}.\end{aligned}$$

Since $B_1 = (a+\delta^{-1})^{\frac{1}{2}}(A_s - (X_1)_s)$, it follows from (4.4.6) and (4.4.7) that

$$\begin{aligned}\|\lambda(B_1)\|^2 &= (1+\delta^{-1})\|F_{\tilde{\mathfrak{A}}}\lambda(A) - F_{\tilde{\mathfrak{A}}}\lambda(X_1)\|^2 \\ &< (1+\delta^{-1})\varepsilon^2 < \frac{\varepsilon}{2^5}.\end{aligned}$$

We apply the similar argument to the element B_1 of $\tau_{F_{\tilde{\mathfrak{A}}}}(\tilde{\mathfrak{A}})$ instead of A_s.

Step 3 There exist a second approximation $A_2 \in \mathfrak{A}$ of B_1 and the rest $B_2 \in \tau_{F_{\tilde{\mathfrak{A}}}}(\tilde{\mathfrak{A}})$ satisfying

(1) $\pi(B_1)^*\pi(B_1) \leqq \pi(A_2)^*\pi(A_2) + \pi(B_2)^*\pi(B_2)$,
(2) $\pi(A_2)^*\pi(A_2) < \frac{\varepsilon}{2^2}$ and $\pi(B_2)^*\pi(B_2) < \frac{\varepsilon}{2^7}$,
(3) $\|F_{\tilde{\mathfrak{A}}}\lambda(A_2)\|^2 < \frac{\varepsilon}{2^2}$ and $\|\lambda(B_2)\|^2 < \frac{\varepsilon}{2^7}$.

Proof

(1) Since $B_1 \in \tau_{F_{\tilde{\mathfrak{A}}}}(\tilde{\mathfrak{A}})$, there exists an element $X_2 \in \mathfrak{A}$ such that

$$\|\pi(B_1) - \pi(X_2)\| < \frac{\sqrt{\varepsilon}}{2^4} \tag{4.4.8}$$

and

$$\|\lambda(B_1) - F_{\tilde{\mathfrak{A}}}\lambda(X_2)\| < \frac{\sqrt{\varepsilon}}{2^4}. \tag{4.4.9}$$

Put

$$A_2 = \sqrt{2}X_2 \quad \text{and} \quad B_2 = \sqrt{2}(B_1 - (X_2)_s).$$

Then, $A_2 \in \mathfrak{A}$ and $B_2 \in \tau_{F_{\tilde{\mathfrak{A}}}}(\tilde{\mathfrak{A}})$. Since $B_1 = (X_2)_s + (B_1 - (X_2)_s)$ and $\pi(X_2) = \pi((X_2)_s)$, we get by (4.4.4) that

$$\begin{aligned}\pi(B_1)^*\pi(B_1) &\leqq 2\pi(X_2)^*\pi(X_2) + 2\pi(B_1 - (X_2)_s)^*\pi(B_1 - (X_2)_s) \\ &= \pi(A_2)^*\pi(A_2) + \pi(B_2)^*\pi(B_2).\end{aligned}$$

(2) Since $X_2 = B_1 + (X_2 - B_1)$, we get by (4.4.4) also that

$$\pi(X_2)^*\pi(X_2) \leqq 2\pi(B_1)^*\pi(B_1) + 2\pi(X_2 - B_1)^*\pi(X_2 - B_1).$$

Hence it follows from Step 1, (2) and (4.4.8) that

$$\begin{aligned}\pi(A_2)^*\pi(A_2) &= 2\pi(X_2)^*\pi(X_2) \\ &\leqq 2^2\pi(B_1)^*\pi(B_1) + 2^2\pi(X_2 - B_1)^*\pi(X_2 - B_1) \\ &< 2^2\frac{\varepsilon}{2^5} + 2^2\frac{\varepsilon}{2^8} < \frac{\varepsilon}{2^2},\end{aligned}$$

and that

$$\pi(B_2)^*\pi(B_2) = 2\pi(B_1 - X_2)^*\pi(B_1 - X_2) < \frac{\varepsilon}{2^7}.$$

(3) Since $B_2 = \sqrt{2}(B_1 - (X_2)_s)$, it follows from (4.4.9) that

$$\begin{aligned}\|\lambda(B_2)\| &= \sqrt{2}\|\lambda(B_1) - F_{\tilde{\mathfrak{A}}}\lambda(X_2)\| \\ &< \sqrt{2}\frac{\sqrt{\varepsilon}}{2^4},\end{aligned}$$

and from Step 2, (3) that

$$\begin{aligned}\|F_{\tilde{\mathfrak{A}}}\lambda(A_2)\| &= \sqrt{2}\|F_{\tilde{\mathfrak{A}}}\lambda(X_2)\| \\ &< \sqrt{2}\left(\|\lambda(B_1)\| + \frac{\sqrt{\varepsilon}}{2^4}\right) \\ &< \sqrt{2}\left(\frac{\sqrt{\varepsilon}}{\sqrt{2}2^2} + \frac{\sqrt{\varepsilon}}{2^4}\right) \\ &< \frac{\sqrt{\varepsilon}}{2}.\end{aligned}$$

Step 4 There exist sequences $\{A_n;\ n = 2, 3, \cdots\}$ in $\mathfrak{A}$ and $\{B_n;\ n = 2, 3, \cdots\}$ in $\tau_{F_{\tilde{\mathfrak{A}}}}(\tilde{\mathfrak{A}})$ which satisfy the following

(1) $\pi(B_{n-1})^*\pi(B_{n-1}) \leqq \pi(A_n)^*\pi(A_n) + \pi(B_n)^*\pi(B_n)$,
(2) $\pi(A_n)^*\pi(A_n) < \frac{\varepsilon}{2^{2(n-1)}}$ and $\pi(B_n)^*\pi(B_n) < \frac{\varepsilon}{2^{2n+3}}$,
(3) $\|F_{\tilde{\mathfrak{A}}}\lambda(A_n)\|^2 < \frac{\varepsilon}{2^{n-1}}$ and $\|\lambda(B_n)\|^2 < \frac{\varepsilon}{2^{2n+3}}$
for $n = 2, 3, \cdots$.

Proof We use the mathematical induction. Step 3 shows that these assertions hold for $n = 2$. Assume that these assertions hold for $k = n$, namely there exist $\{A_k;\ k = 2, \cdots, n\} \subset \mathfrak{A}$ and $\{B_k;\ k = 2, \cdots n\} \subset \tau_{F_{\tilde{\mathfrak{A}}}}(\tilde{\mathfrak{A}})$ satisfying (1) and the following inequalities

$$\pi(A_n)^*\pi(A_n) < \frac{\varepsilon}{2^{2(n-1)}} \quad \text{and} \quad \pi(B_n)^*\pi(B_n) < \frac{\varepsilon}{2^{2n+3}}, \tag{4.4.10}$$

$$\|F_{\tilde{\mathfrak{A}}}\lambda(A_n)\|^2 < \frac{\varepsilon}{2^{n-1}} \quad \text{and} \quad \|\lambda(B_n)\|^2 < \frac{\varepsilon}{2^{2n+3}}. \tag{4.4.11}$$

We will show statements (1), (2) and (3) for $k = n + 1$.
(1) Since $B_n \in \tau_{F_{\tilde{\mathfrak{A}}}}(\tilde{\mathfrak{A}})$, there exists $X_{n+1} \in \mathfrak{A}$ such that $\|B_n - (X_{n+1})_s\| < \frac{\sqrt{\varepsilon}}{2^{n+3}}$, namely

$$\|\pi(B_n) - \pi(X_{n+1})\| < \frac{\sqrt{\varepsilon}}{2^{n+3}}, \tag{4.4.12}$$

$$\|\lambda(B_n) - F_{\tilde{\mathfrak{A}}}\lambda(X_{n+1})\| < \frac{\sqrt{\varepsilon}}{2^{n+3}}. \tag{4.4.13}$$

Using X_{n+1}, we set

$$A_{n+1} = \sqrt{2}X_{n+1} \quad \text{and} \quad B_{n+1} = \sqrt{2}(B_n - (X_{n+1})_s).$$

Then $A_{n+1} \in \mathfrak{A}$ and $B_{n+1} \in \tau_{F_{\tilde{\mathfrak{A}}}}(\tilde{\mathfrak{A}})$. Since $B_n = (X_{n+1})_s + (B_n - (X_{n+1})_s)$, we get by (4.4.4) that

$$\begin{aligned} \pi(B_n)^*\pi(B_n) &\leqq 2\pi(X_{n+1})^*\pi(X_{n+1}) + 2\pi\left(B_n - (X_{n+1})_s\right)^*\pi(B_n - (X_{n+1})_s) \\ &= \pi(A_{n+1})^*\pi(A_{n+1}) + \pi(B_{n+1})^*\pi(B_{n+1}). \end{aligned}$$

(2) Since $X_{n+1} = B_n + (X_{n+1} - B_n)$, we get by (4.4.4) again that

$$\pi(X_{n+1})^*\pi(X_{n+1}) \leqq 2\pi(B_n)^*\pi(B_n) + 2\pi(X_{n+1} - B_n)^*\pi(X_{n+1} - B_n),$$

and by (4.4.10) and (4.4.12) that

$$\begin{aligned}\pi(A_{n+1})^*\pi(A_{n+1}) &= 2\pi(X_{n+1})^*\pi(X_{n+1})\\ &\leqq 2^2\pi(B_n)^*\pi(B_n) + 2^2\pi(X_{n+1}-B_n)^*\pi(X_{n+1}-B_n)\\ &< \frac{\varepsilon}{2^{2n+1}} + \frac{\varepsilon}{2^{2n+4}} < \frac{\varepsilon}{2^{2n}}\end{aligned}$$

and

$$\begin{aligned}\pi(B_{n+1})^*\pi(B_{n+1}) &= 2\pi(B_n - X_{n+1})^*\pi(B_n - X_{n+1})\\ &< 2\frac{\varepsilon}{2^{2n+6}} = \frac{\varepsilon}{2^{2n+5}}.\end{aligned}$$

(3) Since $B_{n+1} = \sqrt{2}(B_n - (X_{n+1})_s)$, it follows from (4.4.13) that

$$\begin{aligned}\|\lambda(B_n)\|^2 &= 2\|\lambda(B_n) - F_{\tilde{\mathfrak{A}}}\lambda(X_{n+1})\|^2\\ &< 2\frac{\varepsilon}{2^{2n+6}} = \frac{\varepsilon}{2^{2n+5}},\end{aligned}$$

which implies that

$$\begin{aligned}\|F_{\tilde{\mathfrak{A}}}\lambda(A_{n+1})\| &= \sqrt{2}\|F_{\tilde{\mathfrak{A}}}\lambda(K_{n+1})\|\\ &< \sqrt{2}(\|\lambda(B_n)\| + \frac{\sqrt{\varepsilon}}{2^{n+3}})\\ &< \left(\frac{1}{2^{n+1}} + \frac{1}{2^{n+2}}\right)\sqrt{\varepsilon} < \frac{\sqrt{\varepsilon}}{2^n}.\end{aligned}$$

Step 5 $\{A_n\} \in \mathfrak{F}(A)$ and $\sum_{n=1}^{\infty} \|F_{\tilde{\mathfrak{A}}}\lambda(A_n)\|^2 < \|F_{\tilde{\mathfrak{A}}}\lambda(A)\|^2 + \varepsilon$.

Proof By Step 2, (1) and Step 4, (1), we have

$$\begin{aligned}\pi(A)^*\pi(A) &\leqq \pi(A_1)^*\pi(A_1) + \pi(B_1)^*\pi(B_1)\\ &\leqq \pi(A_1)^*\pi(A_1) + \big(\pi(A_2)^*\pi(A_2) + \pi(B_2)^*\pi(B_2)\big)\\ &\leqq \pi(A_1)^*\pi(A_1) + \left(\sum_{k=2}^{n}\pi(A_k)^*\pi(A_k) + \pi(B_n)^*\pi(B_n)\right)\\ &= \sum_{k=1}^{n}\pi(A_k)^*\pi(A_k) + \pi(B_n)^*\pi(B_n)\end{aligned}$$

for all $n \in \mathbb{N}$. Since $\lim_{n\to\infty} \|\pi(B_n)\| = 0$ by Step 4, (2), it follows that

$$\pi(A)^*\pi(A) \leqq \sum_{n=1}^{\infty} \pi(A_n)^*\pi(A_n),$$

and from Step 2, (2) and Step 4, (2) that

$$\begin{aligned}\sum_{n=1}^{\infty} \|\pi(A_n)\|^2 &= \|\pi(A_1)\|^2 + \sum_{n=2}^{\infty} \|\pi(A_n)\|^2 \\ &\leqq \left(\|\pi(A)\|^2 + \frac{\varepsilon}{2}\right) + \left(\sum_{n=2}^{\infty} \frac{\varepsilon}{2^{2(n-1)}}\right) \\ &< \|\pi(A)\|^2 + \varepsilon,\end{aligned}$$

which implies that $\{A_n\} \in \mathfrak{F}(A)$. By Step 2, (3) and Step 4, (3) we have

$$\begin{aligned}\sum_{n=1}^{\infty} \|F_{\tilde{\mathfrak{A}}}\lambda(A_n)\|^2 &= \|F_{\tilde{\mathfrak{A}}}\lambda(A)\|^2 + \sum_{n=2}^{\infty} \|F_{\tilde{\mathfrak{A}}}\lambda(A_n)\|^2 \\ &< \|F_{\tilde{\mathfrak{A}}}\lambda(A)\|^2 + \frac{\varepsilon}{2} + \sum_{n=2}^{\infty} \frac{\varepsilon}{2^{n-1}} \\ &= \|F_{\tilde{\mathfrak{A}}}\lambda(A)\|^2 + \varepsilon.\end{aligned}$$

Thus the proof of Theorem 4.4.2 completes from Steps 1 and 4.

4.4.3 *Construction of Semisimple CT^*-Algebras from a T^*-Algebra*

In this subsection we construct semisimple CT^*-algebras from a T^*-algebra. Let $\mathfrak{A}$ be a T^*-algebra on $\mathcal{H}$ and denote the uniform closure $\overline{\pi(\mathfrak{A})[\tau_u^\pi]}$ of $\pi(\mathfrak{A})$ by $\mathfrak{A}_0$ for simplicity. We now write

$$\begin{aligned}\mathfrak{M}_{\mathfrak{A}}^{\pi} = \Big\{\{g_\alpha\} \subset \mathcal{H};\quad & (\pi(A)g_\alpha | g_\beta) = 0, \quad \alpha \neq \beta \\ & \text{and} \quad \sum_{\alpha} \|\pi(A)g_\alpha\|^2 < \infty \quad \text{for all} \quad A \in \mathfrak{A}\Big\},\end{aligned}$$

and for $\{g_\alpha\} \in \mathfrak{M}_{\mathfrak{A}}^{\pi}$,

$$\mathfrak{A}_0(\{g_\alpha\}) = \left\{X_0 \in \mathfrak{A}_0;\ \sum_\alpha \|X_0 g_\alpha\|^2 < \infty \ \text{ and } \ \sum_\alpha \|X_0^* g_\alpha\|^2 < \infty\right\}.$$

Then $\mathfrak{A}_0(\{g_\alpha\})$ is a dense $*$-subalgebra of the C^*-algebra $\mathfrak{A}_0$ containing $\pi(\mathfrak{A})$. We obtain the following main result of this subsection:

Theorem 4.4.3 *Let $\mathfrak{A}$ be a T^*-algebra on $\mathcal{H}$. For any $\{g_\alpha\} \in \mathfrak{M}_{\mathfrak{A}}^{\pi}$, put*

$$\mathfrak{A}(\{g_\alpha\}) = \left\{(\pi(A), \sum_\alpha \pi(A)g_\alpha, (\sum_\alpha \pi(A)^* g_\alpha)^*);\quad A \in \mathfrak{A}\right\}$$

and

$$\tau_{\{g_\alpha\}}(\mathfrak{A}_0) = \left\{(X_0, \sum_\alpha X_0 g_\alpha, (\sum_\alpha X_0^* g_\alpha)^*);\quad X_0 \in \mathfrak{A}_0(\{g_\alpha\})\right\}.$$

Then $\mathfrak{A}(\{g_\alpha\})$ is a semisimple T^*-algebra on $\mathcal{H}$ which is uniformly dense in the semisimple CT^*-algebra $\tau_{\{g_\alpha\}}(\mathfrak{A}_0)$.

Proof It is easily proved that $\tau_{\{g_\alpha\}}(\mathfrak{A}_0)$ is a T^*-algebra on $\mathcal{H}$ containing the T^*-algebra $\mathfrak{A}(\{g_\alpha\})$. We show that $\tau_{\{g_\alpha\}}(\mathfrak{A}_0)$ is a CT^*-algebra on $\mathcal{H}$. Since $(\pi(A)g_\alpha|g_\beta) = 0, \alpha \neq \beta$ for all $A \in \mathfrak{A}$, we have

$$(X_0 g_\alpha|g_\beta) = 0, \quad \alpha \neq \beta \quad \text{for all} \quad X_0 \in \mathfrak{A}_0. \tag{4.4.14}$$

Suppose that $\left\{X_n := (X_0^{(n)}, \sum_\alpha X_0^{(n)} g_\alpha, (\sum_\alpha (X_0^{(n)})^* g_\alpha)^*)\right\}$ is a sequence in $\tau_{\{g_\alpha\}}(\mathfrak{A}_0)$ which converges uniformly to (X_0, x, y^*) in $T^*(\mathcal{H})$. Then $\{X_0^{(n)}\}$ converges uniformly to X_0; hence

$$X_0 \in \mathfrak{A}_0. \tag{4.4.15}$$

Since $\sum_\alpha \|X_0^{(n)} g_\alpha\|^2 < \infty$ for all $n \in \mathbb{N}$, $\Lambda := \cup_{n=1}^{\infty} \left\{\alpha;\ \ X_0^{(n)} g_\alpha \neq 0\right\}$ is a countable set, and so denote it by $\Lambda = \{\alpha_1, \alpha_2, \cdots\}$. Then we have

$$\sum_\alpha \|X_0^{(n)} g_\alpha\|^2 = \sum_{k=1}^{\infty} \|X_0^{(n)} g_{\alpha_k}\|^2 \tag{4.4.16}$$

for all $n \in \mathbb{N}$. For any $\alpha \notin \Lambda$ we get

$$X_0 g_\alpha = \lim_{n\to\infty} X_0^{(n)} g_\alpha = 0.$$

Hence, $\Lambda_0 := \{\alpha; X_0 g_\alpha \neq 0\} \subset \Lambda$, and

$$\sum_\alpha \|X_0 g_\alpha\|^2 = \sum_{k=1}^\infty \|X_0 g_{\alpha_k}\|^2. \tag{4.4.17}$$

Then, it follows from (4.4.14) and (4.4.16) that for any $\varepsilon > 0$ there exists $N \in \mathbb{N}$ such that

$$\begin{aligned} \|\sum_\alpha X_0^{(n)} g_\alpha - \sum_\alpha X_0^{(m)} g_\alpha\|^2 &= \sum_\alpha \|X_0^{(n)} g_\alpha - X_0^{(m)} g_\alpha\|^2 \\ &= \sum_{k=1}^\infty \|X_0^{(n)} g_{\alpha_k} - X_0^{(m)} g_{\alpha_k}\|^2 \\ &< \varepsilon \end{aligned} \tag{4.4.18}$$

for all $n,\ m \geqq N$, and for any $M \in \mathbb{N}$

$$\begin{aligned} \sum_{k=1}^M \|X_0^{(n)} g_{\alpha_k} - X_0 g_{\alpha_k}\|^2 &= \lim_{m\to\infty} \sum_{k=1}^M \|X_0^{(n)} g_{\alpha_k} - X_0^{(m)} g_{\alpha_k}\|^2 \\ &\leqq \varepsilon. \end{aligned}$$

Hence we get

$$\sum_{k=1}^\infty \|X_0^{(n)} g_{\alpha_k} - X_0 g_{\alpha_k}\|^2 \leqq \varepsilon$$

for all $n \geqq N$, which implies that

$$\sum_{k=1}^\infty \|X_0 g_{\alpha_k}\|^2 < \infty \quad \text{and} \quad \lim_{n\to\infty} \|\sum_{k=1}^\infty X_0^{(n)} g_{\alpha_k} - \sum_{k=1}^\infty X_0 g_{\alpha_k}\|^2 = 0,$$

and by (4.4.17) and (4.4.18) that

$$\sum_\alpha \|X_0 g_\alpha\|^2 < \infty \quad \text{and} \quad \lim_{n\to\infty} \|\sum_\alpha X_0^{(n)} g_\alpha - \sum_\alpha X_0 g_\alpha\|^2 = 0.$$

Similarly we can prove that

$$\sum_\alpha \|X_0^* g_\alpha\|^2 < \infty \quad \text{and} \quad \lim_{n\to\infty} \|\sum_\alpha (X_0^{(n)})^* g_\alpha - \sum_\alpha X_0^* g_\alpha\|^2 = 0,$$

which implies by (4.4.15) that $(X_0, \sum_\alpha X_0 g_\alpha, (\sum_\alpha X_0^* g_\alpha)^*) \in \tau_{\{g_\alpha\}}(\mathfrak{A}_0)$. Thus, $\tau_{\{g_\alpha\}}(\mathfrak{A}_0)$ is a CT^*-algebra on $\mathcal{H}$ and is semisimple by Theorem 4.2.6. We will be able to show by Corollary 4.4.13 in Sect. 4.4.5 later, that $\mathfrak{A}(\{g_\alpha\})$ is uniformly dense in the semisimple CT^*-algebra $\tau_{\{g_\alpha\}}(\mathfrak{A}_0)$. Hence, it follows from (4.4.1) that $\mathfrak{A}(\{g_\alpha\})$ is semisimple. This completes the proof.

By Theorem 4.4.3 we can construct a family $\{\mathfrak{A}(\{g_\alpha\});\ \{g_\alpha\} \in \mathfrak{M}_{\mathfrak{A}}^\pi\}$ of semisimple T^*-algebras and a family $\{\tau_{\{g_\alpha\}}(\mathfrak{A}_0);\quad \{g_\alpha\} \in \mathfrak{M}_{\mathfrak{A}}^\pi\}$ of semisimple CT^*-algebras. In Theorem 4.2.6 we have obtained that if $\mathfrak{A}$ is a semisimple CT^*-algebra, then

$$\mathfrak{A} = \mathfrak{A}(\{g_\alpha\}) = \{(\pi(A), \sum_\alpha \pi(A) g_\alpha, (\sum_\alpha \pi(A)^* g_\alpha)^*);\quad A \in \mathfrak{A}\}$$

for some $\{g_\alpha\} \in \mathfrak{M}_{\mathfrak{A}}$. We now consider whether semisimple T^*-algebras satisfy this property. For that, we prepare the following

Lemma 4.4.4 *Let $\mathfrak{A}$ be a T^*-algebra on $\mathcal{H}$, and $\mathfrak{M}_{\mathfrak{A}}$ the set defined already in* (4.2.2) *for $\mathfrak{A}$. Then*

$$\mathfrak{M}_{\mathfrak{A}} = \mathfrak{M}_{\tilde{\mathfrak{A}}} \subset \mathfrak{M}_{\tilde{\mathfrak{A}}}^\pi \subset \mathfrak{M}_{\mathfrak{A}}^\pi.$$

Proof Clearly, $\mathfrak{M}_{\tilde{\mathfrak{A}}} \subset \mathfrak{M}_{\mathfrak{A}} \subset \mathfrak{M}_{\mathfrak{A}}^\pi$. Hence it is sufficient to show $\mathfrak{M}_{\mathfrak{A}} \subset \mathfrak{M}_{\tilde{\mathfrak{A}}}$. Take an arbitrary $\{g_\alpha\} \in \mathfrak{M}_{\mathfrak{A}}$ and $A \in \tilde{\mathfrak{A}}$. Then, choose a sequence $\{A_n\}$ in $\mathfrak{A}$ which converges uniformly to A. Then we can prove in the same way as (4.4.16) and (4.4.17) that

$$\sum_\alpha \|\pi(A_n) g_\alpha\|^2 = \sum_{k=1}^\infty \|\pi(A_n) g_{\alpha_k}\|^2$$

for all $n \in \mathbb{N}$, and that

$$\sum_\alpha \|\pi(A) g_\alpha\|^2 = \sum_{k=1}^\infty \|\pi(A) g_{\alpha_k}\|^2 \tag{4.4.19}$$

for some sequence $\{\alpha_k\}$ in $\{\alpha\}$. Since $\lim_{n\to\infty} \lambda(A_n) = \lambda(A)$, for any $\varepsilon > 0$ there exists $N \in \mathbb{N}$ such that

$$\begin{aligned}\sum_{k=1}^\infty \|\pi(A_n) g_{\alpha_k}\|^2 &\leqq \|\lambda(A_n)\|^2 \\ &\leqq \|\lambda(A)\|^2 + \varepsilon\end{aligned}$$

for all $n \geqq N$; hence for any $M \in \mathbb{N}$

$$\sum_{k=1}^{M} \|\pi(A)g_{\alpha_k}\|^2 = \lim_{n\to\infty} \sum_{k=1}^{M} \|\pi(A_n)g_{\alpha_k}\|^2$$
$$\leqq \|\lambda(A)\|^2 + \varepsilon.$$

Hence it follows from (4.4.19) that

$$\sum_{\alpha} \|\pi(A)g_\alpha\|^2 = \sum_{k=1}^{\infty} \|\pi(A)g_{\alpha_k}\|^2 \leqq \|\lambda(A)\|^2,$$

which means that $A \in \mathfrak{M}_{\tilde{\mathfrak{A}}}$. Thus, $\mathfrak{M}_{\mathfrak{A}} \subset \mathfrak{M}_{\tilde{\mathfrak{A}}}$. This completes the proof.

For semisimple T^*-algebras we have the following

Proposition 4.4.5 *Suppose that $\mathfrak{A}$ is a semisimple T^*-algebra on $\mathcal{H}$. Then there exists a $\{g_\alpha\} \in \mathfrak{M}_{\mathfrak{A}}$ such that $\mathfrak{A} = \mathfrak{A}(\{g_\alpha\})$ and $\tilde{\mathfrak{A}} = \tilde{\mathfrak{A}}(\{g_\alpha\}) = \tau_{\{g_\alpha\}}(\mathfrak{A}_0)$.*

Proof Since $\tilde{\mathfrak{A}}$ is a semisimple CT^*-algebra on $\mathcal{H}$, it follows from Theorem 4.2.6 and Lemma 4.4.4 that $\tilde{\mathfrak{A}} = \tilde{\mathfrak{A}}(\{g_\alpha\})$ for some $\{g_\alpha\} \in \mathfrak{M}_{\mathfrak{A}}$, which implies that $\mathfrak{A} = \mathfrak{A}(\{g_\alpha\})$, and by Theorem 4.4.3 it is uniformly dense in the CT^*-algebra $\tau_{\{g_\alpha\}}(\mathfrak{A}_0)$. Hence, $\tilde{\mathfrak{A}} = \tilde{\mathfrak{A}}(\{g_\alpha\}) = \tau_{\{g_\alpha\}}(\mathfrak{A}_0)$. This completes the proof.

4.4.4 Semisimplicity and Singularity of T^*-Algebras

In Sect. 4.2 we have discussed the semisimplicity and the singularity of CT^*-algebras. In this subsection we consider the case of T^*-algebras. For the semisimplicity of T^*-algebras we have the following

Theorem 4.4.6 *Let $\mathfrak{A}$ be a T^*-algebra on $\mathcal{H}$ and $\tilde{\mathfrak{A}}$ the uniform closure of $\mathfrak{A}$. Then the following statements are equivalent:*

(i) $\mathfrak{A}$ is semisimple.
(ii) There exists an element $\{g_\alpha\}$ of $\mathfrak{M}_{\mathfrak{A}}$ such that

$$\lambda(A) = \sum_{\alpha} \pi(A)g_\alpha \quad \text{for all} \quad A \in \mathfrak{A}.$$

(iii) $\tilde{\mathfrak{A}}$ is semisimple.
(iv) There exists an element $\{g_\alpha\}$ of $\mathfrak{M}_{\mathfrak{A}}$ such that

$$\lambda(A) = \sum_{\alpha} \pi(A)g_\alpha \quad \text{for all} \quad A \in \tilde{\mathfrak{A}}.$$

Proof This follows from Theorem 4.2.6 and Proposition 4.4.5.

To consider the singularity of $\mathfrak{A}$ we define the following set $s(\mathfrak{A})$ by

$$s(\mathfrak{A}) = \{\{A_n\} \subset \mathfrak{A};\quad \pi(A_n) \to 0 \quad \text{uniformly}, \\ \text{and} \quad \{\lambda(A_n)\} \text{ and } \{\lambda(A_n^\sharp)\} \text{ are Cauchy sequences in } \mathcal{H}\}.$$

Then we have the following

Theorem 4.4.7 *Let $\mathfrak{A}$ be a T^*-algebra on $\mathcal{H}$ and $\tilde{\mathfrak{A}}$ the uniform closure of $\mathfrak{A}$. Then the following statements (1), (2) and (3) hold:*

(1) The following are equivalent:

- *(i) $\mathfrak{A}$ is singular.*
- *(ii) $\tilde{\mathfrak{A}}$ is singular.*
- *(iii) $\lambda(s(\mathfrak{A})) := \{\lim_{n\to\infty} \lambda(A_n);\quad \{A_n\} \in s(\mathfrak{A})\}$ is dense in $\mathcal{H}$.*

(2) The following are equivalent:

- *(i) $\mathfrak{A}$ is nondegenerate and singular.*
- *(ii) $\mathfrak{A}^2$ is uniformly dense in $N(T^*(\mathcal{H}))$. Here $\mathfrak{A}^2$ is the $*$-subalgebra of $\mathfrak{A}$ spanned by $\{AB;\ A, B \in \mathfrak{A}\}$.*

(3) The following are equivalent:

- *(i) $\mathfrak{A}$ is nilpotent.*
- *(ii) $\mathfrak{A} = N(\mathfrak{A})$.*
- *(iii) $\tilde{\mathfrak{A}}$ is nilpotent.*
- *(iv) $\tilde{\mathfrak{A}} = N(\tilde{\mathfrak{A}})$.*

Proof

(1) By the definition of singular T^*-algebras (Definition 4.4.1), $\mathfrak{A}$ is singular if and only if $\tilde{\mathfrak{A}}$ is singular if and only if $N_{\tilde{\mathfrak{A}}} = I$. Furthermore, we can prove immediately that $\lambda(N(\tilde{\mathfrak{A}})) = \lambda(s(\mathfrak{A}))$, which yields that (ii) and (iii) are equivalent.

(2) (i)$\Rightarrow$(ii) By assumption (i), $\tilde{\mathfrak{A}}$ is nondegenerate and singular. Hence it follows from Theorem 4.2.9 that for any $x,\ y \in \mathcal{H}$ there exists an element A of $N(\tilde{\mathfrak{A}})$ such that $\lambda(A) = x$ and $\lambda(A^\sharp) = y$, which implies that there exists a sequence $\{A_n\}$ in $\mathfrak{A}$ such that $A_n = (\pi(A_n), \lambda(A_n), \lambda(A_n^\sharp)^*) \to (0, x, y^*)$, uniformly. Furthermore, since $P_{\mathfrak{A}} = I$, we can find a net $\{e_\alpha\}$ in $u_\pi(\mathfrak{A})$ such that $\pi(e_\alpha) \to I$, strongly. Thus, we obtain that $e_\alpha A_n + A_n e_\alpha^\sharp \in \mathfrak{A}^2$ for all $n \in \mathbb{N}$ and α, and that

$$\begin{aligned}\lim_\alpha \lim_{n\to\infty} (e_\alpha A_n + A_n e_\alpha^\sharp) &= \lim_\alpha (0, \pi(e_\alpha)x, (\pi(e_\alpha)^* y)^*) \\ &= (0, x, y^*), \quad \text{uniformly.}\end{aligned}$$

Hence, $\mathfrak{A}^2$ is uniformly dense in $N(T^*(\mathcal{H}))$.

(ii)⇒(i) By assumption (ii), we get $\lambda(N(\tilde{\mathfrak{A}})) = \mathcal{H}$. Hence, $\tilde{\mathfrak{A}}$ is singular, so $\mathfrak{A}$ is also singular. Furthermore, since $[\pi(\mathfrak{A})\lambda(\mathfrak{A})] = [\lambda(\mathfrak{A}^2)]$ is dense in $\mathcal{H}$, it follows that $\mathfrak{A}$ is nondegenerate.

(3) The equivalence of (i), (iii) and (iv) is trivial. Furthermore, we can immediately prove that (ii) and (iv) are equivalent. This completes the proof.

4.4.5 A Natural Weight on the von Neumann Algebra Defined from a T^*-Algebra

Let $\mathfrak{A}$ be a T^*-algebra on $\mathcal{H}$, and denote by $\mathfrak{M}_0$ the von Neumann algebra $\pi(\mathfrak{A})''$. In this subsection we define and investigate a weight on the positive cone $(\mathfrak{M}_0)_+$ of the von Neumann algebra $\mathfrak{M}_0$. This weight plays an important role in next Sect. 4.4.6.

Lemma 4.4.8 *Put*

$$\varphi(X_0) = \sup\left\{(X_0\lambda(K)|\lambda(K));\quad K \in u_\pi(\mathfrak{A}^\tau)\right\},\quad X_0 \in (\mathfrak{M}_0)_+.$$

Then, for $X_0 \in (\mathfrak{M}_0)_+$ we have

$$\varphi(X_0) < \infty \quad \text{if and only if} \quad X_0^{\frac{1}{2}} \in \pi(\mathfrak{A}^{\tau\tau}).$$

Proof Suppose that $\varphi(X_0) < \infty$. Then there exists a positive constant $\gamma > 0$ such that

$$\|X_0^{\frac{1}{2}}\lambda(K)\| \leqq \gamma\|\pi(K)\| \tag{4.4.20}$$

for all $K \in \mathfrak{A}^\tau$. We can prove in the same way as the proof of Lemma 2.2.1 that $X_0^{\frac{1}{2}} \in \pi(\mathfrak{A}^{\tau\tau})$. Indeed, for arbitrary $\varepsilon > 0$ and $\{K_i\}_{i=1,2,\cdots,n} \subset \mathfrak{A}^\tau$ we put

$$\Delta(\varepsilon, \{K_i\}) = \{g \in \mathcal{H};\quad \|g\| \leqq \gamma$$
$$\text{and}\quad \|X_0^{\frac{1}{2}}\lambda(K_i) - \pi(K_i)g\| \leqq \gamma\varepsilon\}.$$

Then, since the C^*-algebra $\overline{\pi(\mathfrak{A}^\tau)[\tau_u^\pi]}$ has an approximate identity, we can take an element K of $\mathfrak{A}^\tau$ such that

$$\|\pi(K)\| \leqq 1 \quad \text{and} \quad \|\pi(K_i)\pi(K) - \pi(K_i)\| < \varepsilon$$

for $i = 1, 2, \cdots, n$, which implies by (4.2.20) that

$$\|X_0^{\frac{1}{2}}\lambda(K)\| \leqq \gamma\|\pi(K)\| \leqq \gamma,$$

$$\begin{aligned}\|X_0^{\frac{1}{2}}\lambda(K_i) - \pi(K_i)X_0^{\frac{1}{2}}\lambda(K)\| &= \|X_0^{\frac{1}{2}}\lambda(K_i - K_iK)\| \\ &\leqq \gamma\|\pi(K_i) - \pi(K_i)\pi(K)\| \\ &\leqq \gamma\varepsilon, \quad i = 1, 2, \cdots, n.\end{aligned}$$

Hence, $X_0^{\frac{1}{2}}\lambda(K) \in \Delta(\varepsilon, \{K_i\})$. Thus, we see that $\Delta(\varepsilon, \{K_i\})$ is a nonempty weakly compact subset of $\mathcal{H}$, so that $\cap\{\Delta(\varepsilon, K); \quad \varepsilon > 0, \quad K \dot{\in} \mathfrak{A}^\tau\} \neq \emptyset$, and choose an element g in it. Then we have

$$X_0^{\frac{1}{2}}\lambda(K) = \pi(K)g$$

for all $K \in \mathfrak{A}^\tau$. Put $X = (X_0^{\frac{1}{2}}, g, g^*)$. Then

$$X \in \mathfrak{A}^{\tau\tau}, \quad \text{so} \quad X_0^{\frac{1}{2}} = \pi(X) \in \pi(\mathfrak{A}^{\tau\tau}).$$

Conversely, suppose that $X = (X_0^{\frac{1}{2}}, g, h^*) \in \mathfrak{A}^{\tau\tau}$. Then we have

$$\begin{aligned}\varphi(X_0) &= \sup\{(X_0\lambda(K)|\lambda(K)); \quad K \in u_\pi(\mathfrak{A}^\tau)\} \\ &= \sup\{\|X_0^{\frac{1}{2}}\lambda(K)\|^2; \quad K \in u_\pi(\mathfrak{A}^\tau)\} \\ &= \sup\{\|\pi(K)g\|^2; \quad K \in u_\pi(\mathfrak{A}^\tau)\} \\ &\leqq \|g\|^2 < \infty,\end{aligned}$$

which completes the proof.

Lemma 4.4.9 *Suppose that $\mathfrak{A}$ is a semisimple T^*-algebra on $\mathcal{H}$. Then,*

$$\pi(\mathfrak{A}^\tau)'' = \pi(\mathfrak{A})' \quad \textit{and} \quad \pi(\mathfrak{A}^{\tau\tau}) \subset \pi(\mathfrak{A})'' = \pi(\mathfrak{A}^{\tau\tau})''.$$

Proof Since $\pi(\mathfrak{A}^\tau) \subset \pi(\mathfrak{A})'$, we see that $\pi(\mathfrak{A}^\tau)'' \subset \pi(\mathfrak{A})'$. Take an arbitrary $K_0 \in \pi(\mathfrak{A})'$, and choose $K \in T^*(\mathcal{H})$ with $\pi(K) = K_0$. Since $\pi(\mathfrak{A}^\tau)$ is a nondegenerate $*$-subalgebra of $\pi(\mathfrak{A})'$ by the semisimplicity of $\mathfrak{A}$, we can take a net $\{e'_\gamma\}$ in $u_\pi(\mathfrak{A}^\tau)$ such that $\pi(e'_\gamma)$ converges strongly to I. Then we have

$$e'_\gamma K e'_\gamma \in \mathfrak{A}^\tau \quad \text{and} \quad \pi(e'_\gamma)K_0\pi(e'_\gamma) = \pi(e'_\gamma K e'_\gamma) \longrightarrow K_0, \quad \text{weakly}.$$

Hence, $K_0 \in \pi(\mathfrak{A}^\tau)''$. Thus, $\pi(\mathfrak{A}^\tau)'' = \pi(\mathfrak{A})'$. Furthermore, since $\pi(\mathfrak{A}) \subset \pi(\mathfrak{A}^{\tau\tau}) \subset \pi(\mathfrak{A})''$, we have $\pi(\mathfrak{A}^{\tau\tau})'' = \pi(\mathfrak{A})''$. This completes the proof.

For a CT^*-algebra $\mathfrak{A}$ we define the π-positive cone $\mathfrak{A}_+^\pi$ of $\mathfrak{A}$ by

$$\mathfrak{A}_+^\pi = \{A \in \mathfrak{A};\quad \pi(A) \geqq 0\}.$$

By Theorem 2.3.2, the square root $A^{\frac{1}{2}}$ of $A \in \mathfrak{A}_+^\pi$ can be defined in $CT^*(A)$. Now we obtain the following main result of this subsection.

Theorem 4.4.10 *Suppose that $\mathfrak{A}$ is a semisimple T^*-algebra on $\mathcal{H}$. Then, the map φ defined on $(\mathfrak{M}_0)_+$ in Lemma 4.4.8 satisfies the following properties:*

(i) $\varphi(\pi(A)) = \|\lambda(A^{\frac{1}{2}})\|^2$ *for all* $A \in u_\pi((\mathfrak{A}^{\tau\tau})_+^\pi)$.
(ii) $M_0(\varphi) := \{X_0 \in \mathfrak{M}_0;\quad \varphi(X_0^* X_0) < \infty \text{ and } \varphi(X_0 X_0^*) < \infty\} = \pi(\mathfrak{A}^{\tau\tau})$.
(iii) φ *is a weight on* $(\mathfrak{M}_0)_+$, *that is, it is a map from* $(\mathfrak{M}_0)_+$ *to* $[0, \infty]$ *satisfying*

$$\varphi(X_0 + Y_0) = \varphi(X_0) + \varphi(Y_0),$$
$$\varphi(\alpha X_0) = \alpha\varphi(X_0)$$

for all $X_0,\ Y_0 \in (\mathfrak{M}_0)_+$ *and* $\alpha \geqq 0$, *where* $0 \cdot \infty := 0$.
(iv) $\varphi(X_0) = \sup\{\varphi(\pi(A));\ A \in \mathfrak{A}^{\tau\tau} \text{ with } 0 \leqq \pi(A) \leqq X_0\},\ X_0 \in (\mathfrak{M}_0)_+$.

Proof

(i) Suppose that $A \in (\mathfrak{A}^{\tau\tau})_+^\pi$. Since $\pi(\mathfrak{A}^\tau)$ is nondegenerate, we have

$$\begin{aligned}
\varphi(\pi(A)) &= \sup\{(\pi(A)\lambda(K)|\lambda(K));\quad K \in u_\pi(\mathfrak{A}^\tau)\}\\
&= \sup\{\|\pi(A^{\frac{1}{2}})\lambda(K)\|^2;\quad K \in u_\pi(\mathfrak{A}^\tau)\}\\
&= \sup\{\|\pi(K)\lambda(A^{\frac{1}{2}})\|^2;\quad K \in u_\pi(\mathfrak{A}^\tau)\}\\
&= \|\lambda(A^{\frac{1}{2}})\|^2.
\end{aligned}$$

(ii) Let $X_0 \in M_0(\varphi)$ and $X_0 = U|X_0|$ be the polar decomposition of X_0. By Lemma 4.4.8 both $|X_0| = (X_0^* X_0)^{\frac{1}{2}}$ and $|X_0^*| = (X_0 X_0^*)^{\frac{1}{2}}$ belong to $\pi(\mathfrak{A}^{\tau\tau})$. Hence we can take elements x, y, z, w of $\mathcal{H}$ such that $(|X_0|, x, y^*)$ and $(|X_0^*|, z, w^*)$ belong to $\mathfrak{A}^{\tau\tau}$. Since $X_0 = U|X_0| = |X_0^*|U$, we can show that $(X_0, Ux, (U^*w)^*) \in \mathfrak{A}^{\tau\tau}$, which yields that $X_0 \in \pi(\mathfrak{A}^{\tau\tau})$ and $M_0(\varphi) \subset \pi(\mathfrak{A}^{\tau\tau})$. Conversely, take an arbitrary $A \in \mathfrak{A}^{\tau\tau}$. Since $(A^\sharp A)^{\frac{1}{2}} \in CT^*(A) \subset \mathfrak{A}^{\tau\tau}$ by Theorem 2.3.7, it follows from Lemma 4.4.8 that $\pi(A) \in M_0(\varphi)$, and $\pi(\mathfrak{A}^{\tau\tau}) \subset M_0(\varphi)$. Thus, $M_0(\varphi) = \pi(\mathfrak{A}^{\tau\tau})$.
(iii) Let X_0, Y_0 in $(\mathfrak{M}_0)_+$. It is clear that

$$\varphi(X_0 + Y_0) \leqq \varphi(X_0) + \varphi(Y_0).$$

Therefore, it suffices to show that

$$\varphi(X_0) + \varphi(Y_0) \leqq \varphi(X_0 + Y_0). \tag{4.4.21}$$

In case that $\varphi(X_0) = \infty$ or $\varphi(Y_0) = \infty$, this follows from the inequalities:

$$\varphi(X_0) \leqq \varphi(X_0 + Y_0) \quad \text{and} \quad \varphi(Y_0) \leqq \varphi(X_0 + Y_0).$$

Suppose that $\varphi(X_0) < \infty$ and $\varphi(Y_0) < \infty$. By Lemma 4.4.8 there exist X, $Y \in \mathfrak{A}^{\tau\tau}$ such that $\pi(X) = X_0^{\frac{1}{2}}$ and $\pi(Y) = Y_0^{\frac{1}{2}}$, and by (i)

$$\varphi(X_0) = \|\lambda(X)\|^2 \quad \text{and} \quad \varphi(Y_0) = \|\lambda(Y)\|^{\frac{1}{2}}.$$

Hence we have

$$\begin{aligned} \varphi(X_0) + \varphi(Y_0) &= \|\lambda(X)\|^2 + \|\lambda(Y)\|^2 \\ &= \lim_{\gamma} \left\{ \|\pi(e'_\gamma)\lambda(X)\|^2 + \|\pi(e'_\gamma)\lambda(Y)\|^2 \right\} \\ &= \lim_{\gamma} \left((X_0 + Y_0)\lambda(e'_\gamma) | \lambda(e'_\gamma) \right) \\ &\leqq \varphi(X_0 + Y_0), \end{aligned}$$

where $\{e'_\gamma\}$ is a net in $u_\pi(\mathfrak{A}^\tau)$ such that $\pi(e'_\gamma)$ converges strongly to I, which implies (4.4.21), and that $\varphi(X_0 + Y_0) = \varphi(X_0) + \varphi(Y_0)$. Since $\pi(\mathfrak{A}^{\tau\tau}) \subset \pi(\mathfrak{A})'' = \mathfrak{M}_0$, we see that $\varphi(\alpha X_0) = \alpha\varphi(X_0)$ for all $X_0 \in (\mathfrak{M}_0)_+$ and $\alpha \geqq 0$.

(iv) Let $X_0 \in (\mathfrak{M}_0)_+$. Since $\pi(\mathfrak{A}^{\tau\tau}) \subset \pi(\mathfrak{A})'' = \mathfrak{M}_0$, we see that

$$\sup\{\varphi(\pi(A)); \quad A \in \mathfrak{A}^{\tau\tau} \quad \text{with} \quad 0 \leqq \pi(A) \leqq X_0\} \leqq \varphi(X_0),$$

and that the equality holds in case that $X_0 \in \pi(\mathfrak{A}^{\tau\tau})$. Suppose $X_0 \notin \pi(\mathfrak{A}^{\tau\tau})$. By Lemma 4.4.8 we have

$$\varphi(X_0) = \infty. \tag{4.4.22}$$

Since $\pi(\mathfrak{A}^{\tau\tau})'' = \pi(\mathfrak{A})''$ by Lemma 4.4.9, there exists a net $\{A_\alpha\}$ in $(\mathfrak{A}^{\tau\tau})^\pi_+$ such that $\{\pi(A_\alpha\}$ converges strongly to X_0. Then, since

$$\begin{aligned} (X_0\lambda(K)|\lambda(K)) &= \lim_{\alpha}(\pi(A_\alpha)\lambda(K)|\lambda(K)) \\ &\leqq \underline{\lim}_\alpha \varphi(\pi(A_\alpha)) \end{aligned}$$

for all $K \in u_\pi(\mathfrak{A}^\tau)$, we have

$$\begin{aligned}\varphi(X_0) &= \sup\{(X_0\lambda(K)|\lambda(K)); \quad K \in u_\pi(\mathfrak{A}^\tau)\} \\ &\leqq \underline{\lim}_\alpha \varphi(\pi(A_\alpha)).\end{aligned}$$

Hence it follows from (4.4.22) that $\underline{\lim}_\alpha \varphi(\pi(A_\alpha)) = \infty$, which implies that

$$\sup\left\{\varphi(\pi(A)); \quad A \in \mathfrak{A}^{\tau\tau} \quad \text{with} \quad 0 \leqq \pi(A) \leqq X_0\right\} = \infty.$$

Thus, (iv) holds. This completes the proof.

Corollary 4.4.11 *Suppose that $\mathfrak{A}$ is a regular WT^*-algebra on $\mathcal{H}$. Then there exists a weight φ on $(\mathfrak{M}_0)_+$ such that*

(i) $M_0(\varphi) = \pi(\mathfrak{A})$,
(ii) $\varphi(\pi(A)) = \|\lambda(A^{\frac{1}{2}})\|^2$ *for all* $A \in \mathfrak{A}^\pi_+$,
(iii) $\varphi(X_0) = \sup\left\{\varphi(\pi(A)); \quad A \in \mathfrak{A} \quad \textit{with} \quad 0 \leqq \pi(A) \leqq X_0\right\}$, $X_0 \in (\mathfrak{M}_0)_+$.

Proof Since $\mathfrak{A}$ is a regular WT^*-algebra, it follows from Theorems 3.1.1 and 4.2.8 that $\mathfrak{A} = \mathfrak{A}^T = \mathfrak{A}^{\tau\tau}$. Therefore, the assertion follows from Theorem 4.4.10.

4.4.6 Density of a ∗-Subalgebra of a T^*-Algebra

We first investigate the density of a ∗-subalgebra of a T^*-algebra under the strong* topology. For that, we prepare the following

Lemma 4.4.12 *Let $\mathfrak{A}$ and $\mathfrak{B}$ be regular WT^*-algebras on $\mathcal{H}$. Suppose that $\mathfrak{B}$ is a ∗-subalgebra of $\mathfrak{A}$ such that $\pi(\mathfrak{B})$ is strongly dense in $\pi(\mathfrak{A})$. Then $\mathfrak{B} = \mathfrak{A}$.*

Proof By Theorem 4.2.9 we have

$$\mathfrak{A}^{\tau\tau} = \mathfrak{A} \quad \text{and} \quad \mathfrak{B}^{\tau\tau} = \mathfrak{B}. \tag{4.4.23}$$

By Corollary 4.4.11 there exist weights φ and ψ on $(\mathfrak{M}_0)_+$, where $\mathfrak{M}_0 := \pi(\mathfrak{A})'' = \pi(\mathfrak{B})''$ such that

$$\begin{cases}\varphi(\pi(A)) = \|\lambda(A^{\frac{1}{2}})\|^2, \quad A \in \mathfrak{A}^\pi_+, \\ M_0(\varphi) = \pi(\mathfrak{A}), \\ \varphi(X_0) = \sup\{\varphi(\pi(A)); \quad A \in \mathfrak{A} \text{ with } 0 \leqq \pi(A) \leqq X_0\}, \quad X_0 \in (\mathfrak{M}_0)_+;\end{cases} \tag{4.4.24}$$

and

$$\begin{cases} \psi(\pi(B)) = \|\lambda(B^{\frac{1}{2}})\|^2, \quad B \in \mathfrak{B}_+^{\pi}, \\ M_0(\psi) = \pi(\mathfrak{B}), \\ \psi(X_0) = \sup\{\psi(\pi(B)); \quad B \in \mathfrak{B} \text{ with } 0 \leqq \pi(B) \leqq X_0\}, \quad X_0 \in (\mathfrak{M}_0)_+. \end{cases} \tag{4.4.25}$$

Take an arbitrary $A \in \mathfrak{A}$. By the regularity of $\mathfrak{A}$ we have

$$A \in \mathfrak{B} \quad \text{if and only if} \quad \pi(A) \in \pi(\mathfrak{B}). \tag{4.4.26}$$

Indeed, if $\pi(A) \in \pi(\mathfrak{B})$, then there exists an element B of $\mathfrak{B}$ such that $B = (\pi(A), \lambda(B), \lambda(B^\sharp)^*)$, so

$$A - B = (0, \lambda(A) - \lambda(B), (\lambda(A^\sharp) - \lambda(B^\sharp))^*) \in N(\mathfrak{A}).$$

Since $\mathfrak{A}$ is regular, we get $A = B \in \mathfrak{B}$. The converse is trivial. Suppose now that $\mathfrak{A} \neq \mathfrak{B}$. Then, for $A \in \mathfrak{A}\backslash\mathfrak{B}$ either its real part $\frac{A+A^\sharp}{2}$ or its imaginary part $\frac{A-A^\sharp}{2i}$ don't belong to $\mathfrak{B}$. Hence, by (4.4.26) there exists an hermitian element H of $\mathfrak{A}$ such that $\pi(H)^* = \pi(H) \notin \pi(\mathfrak{B})$. Let $\pi(H) = U_H|\pi(H)|$ be the polar decomposition of $\pi(H)$. By Theorem 2.3.7 the square root $(H^\sharp H)^{\frac{1}{2}}$ of $H^\sharp H$:

$$(H^\sharp H)^{\frac{1}{2}} = (|\pi(H)|, U_H\lambda(H), (U_H\lambda(H))^*)$$

belongs to $\mathfrak{A}$. Assume now that $(H^\sharp H)^{\frac{1}{2}} \in \mathfrak{B}$. Then, write

$$X = (\pi(H), U_H^2\lambda(H), (U_H^2\lambda(H))^*).$$

Since $\pi(\mathfrak{A})'' = \pi(\mathfrak{B})''$ and $(H^\sharp H)^{\frac{1}{2}} \in \mathfrak{B}$ by assumption, it follows that $U_H, \ \pi(H) \in \pi(\mathfrak{B})''$ and $|\pi(H)|\lambda(K) = \pi(K)U_H\lambda(H) = U_H\pi(K)\lambda(H)$ for all $K \in \mathfrak{B}^\tau$, which implies that

$$\begin{aligned} XK &= (\pi(H)\pi(K), \pi(H)\lambda(K), (\pi(K)^*U_H^2\lambda(H))^*) \\ &= (\pi(K)\pi(H), U_H|\pi(H)|\lambda(K), (U_H\pi(K)^*U_H\lambda(H))^*) \\ &= (\pi(K)\pi(H), U_H^2\pi(K)\lambda(H), (U_H|\pi(H)|\lambda(K^\sharp))^*) \\ &= (\pi(K)\pi(H), \pi(K)U_H^2\lambda(H), (\pi(H)\lambda(K^\sharp))^*) \\ &= KX \end{aligned}$$

for all $K \in \mathfrak{B}^\tau$, so that $X \in \mathfrak{B}^{\tau\tau} = \mathfrak{B}$ by (4.4.22). Hence, $\pi(H) = \pi(X) \in \pi(\mathfrak{B})$, which is a contradiction. Thus $(H^\sharp H)^{\frac{1}{2}} \in \mathfrak{A}\backslash\mathfrak{B}$, which shows that we can assume

$A \in \mathfrak{A}_+^\pi \backslash \mathfrak{B}_+^\pi$ without loss of generality. By (4.4.24) and (4.4.25) we have

$$\varphi(\pi(A)) < \infty \quad \text{and} \quad \psi(\pi(A)) = \infty,$$

and

$$\begin{aligned}\psi(\pi(A)) &= \sup\{\psi(\pi(B)); \quad B \in \mathfrak{B} \text{ with } 0 \leqq \pi(B) \leqq \pi(A)\} \\ &= \sup\{\varphi(\pi(B)); \quad B \in \mathfrak{B} \text{ with } 0 \leqq \pi(B) \leqq \pi(A)\} \\ &\leqq \varphi(\pi(A)) \\ &< \infty.\end{aligned}$$

This is a contradiction. Thus, $\mathfrak{A} = \mathfrak{B}$. This completes the proof.

The next theorem is one of the main results of this subsection.

Theorem 4.4.13 *Let $\mathfrak{A}$ be a T^*-algebra on $\mathcal{H}$ and $\mathfrak{B}$ a $*$-subalgebra of $\mathfrak{A}$. Suppose that*

(i) $\pi(\mathfrak{B})$ is strongly dense in $\pi(\mathfrak{A})$,
(ii) $N(\mathfrak{B})$ is strongly dense in $N(\mathfrak{A})$.

Then $\mathfrak{B}$ is strongly * dense in $\mathfrak{A}$.

Proof Since the strong* closures $\bar{\mathfrak{A}}[\tau_s^*]$ and $\bar{\mathfrak{B}}[\tau_s^*]$ of $\mathfrak{A}$ and $\mathfrak{B}$ satisfy assumptions (i) and (ii), it suffices to show $\mathfrak{A} = \mathfrak{B}$ in case that $\mathfrak{A}$ and $\mathfrak{B}$ are strong* closed, that is, they are WT^*-algebras. Since $N(\mathfrak{A})$ and $N(\mathfrak{B})$ are strongly closed, it follows from assumption (ii) that $N(\mathfrak{A}) = N(\mathfrak{B})$, which implies by (i) that $F_{\mathfrak{A}} = F_{\mathfrak{B}}$. We now denote it by F. Furthermore, since

$$\mathfrak{A} = \tau_F(\mathfrak{A}) + N(\mathfrak{A}) \quad \text{and} \quad \mathfrak{B} = \tau_F(\mathfrak{B}) + N(\mathfrak{B})$$

by Theorem 4.1.6, we have only to show that $\tau_F(\mathfrak{A}) = \tau_F(\mathfrak{B})$, so that we may assume that $\mathfrak{A}$ and $\mathfrak{B}$ are semisimple. Then,it follows from assumption (i) and the semisimplicity of $\mathfrak{A}$ that

$$P := P_{\mathfrak{A}} = P_{\mathfrak{B}} = F, \quad \pi(A)P = \pi(A) \quad \text{and} \quad P\lambda(A) = \lambda(A) \tag{4.4.27}$$

for all $A \in \mathfrak{A}$, which yields that

$$\begin{aligned}\mathfrak{A}_P &:= \left\{(\pi(A)\lceil_{P\mathcal{H}}, \lambda(A), \lambda(A^\sharp)^*); \quad A \in \mathfrak{A}\right\}, \\ \mathfrak{B}_P &:= \left\{(\pi(B)\lceil_{P\mathcal{H}}, \lambda(B), \lambda(B^\sharp)^*); \quad B \in \mathfrak{B}\right\}\end{aligned}$$

are regular WT^*-algebras on $P\mathcal{H}$, and that $\pi(\mathfrak{B}_p)$ is strongly dense in $\pi(\mathfrak{A}_p)$. Hence it follows from Lemma 4.4.12 that $\mathfrak{A}_p = \mathfrak{B}_p$, which implies by (4.4.27) that $\tau_F(\mathfrak{A}) = \tau_F(\mathfrak{B})$. This completes the proof.

The following is an immediate consequence of Theorem 4.4.13.

Corollary 4.4.14 *Let $\mathfrak{A}$ be a T^*-algebra on $\mathcal{H}$ and $\mathfrak{B}$ a $*$-subalgebra of $\mathfrak{A}$. Suppose that $\mathfrak{A}$ is semisimple and $\pi(\mathfrak{B})$ is strongly dense in $\pi(\mathfrak{A})$. Then $\mathfrak{B}$ is strongly* dense in $\mathfrak{A}$.*

Taking notice of the subspaces $\nu(\mathfrak{B})$ and $\nu(\mathfrak{A})$ in the Hilbert space $\mathcal{H} \oplus \mathcal{H}^*$, we have the following

Theorem 4.4.15 *Let $\mathfrak{A}$ be a semisimple T^*-algebra on $\mathcal{H}$ and $\mathfrak{B}$ a $*$-subslgebra of $\mathfrak{A}$. Suppose that the central support $Z(C_{\mathfrak{A}}):=$ Proj $[\pi(\mathfrak{A})'\lambda(\mathfrak{A})]$ of $C_{\mathfrak{A}}$ is the identity I, and $[\nu(\mathfrak{B})] = [\nu(\mathfrak{A})]$ in the Hilbert space $\mathcal{H} \oplus \mathcal{H}^*$. Then, $\mathfrak{B}^\tau = \mathfrak{A}^\tau$ and $\mathfrak{B}$ is σ-strongly * dense in $\mathfrak{A}$.*

Proof We may assume without loss of generality that $\mathfrak{A}$ is a CT^*-algebra. Since $P_{\mathfrak{A}^\tau} = I$ by the semisimplicity of $\mathfrak{A}$, we have

$$\begin{aligned} Z(C_{\mathfrak{A}}) &= \text{Proj } [\pi(\mathfrak{A}^\tau)\lambda(\mathfrak{A})] \\ &= I. \end{aligned} \tag{4.4.28}$$

It is easy to show that $\tilde{\mathfrak{A}}^\tau = \mathfrak{A}^\tau$ and $\tilde{\mathfrak{A}}^\rho = \mathfrak{A}^\rho$, where $\tilde{\mathfrak{A}}$ is the uniform closure of $\mathfrak{A}$. Furthermore, since $\tilde{\mathfrak{A}}$ is a semisimple CT^*-algebra on $\mathcal{H}$, it follows from Theorem 4.1.7 that $\tilde{\mathfrak{A}}^\tau = \tilde{\mathfrak{A}}^\rho$, so that

$$\mathfrak{A}^\tau = \mathfrak{A}^\rho. \tag{4.4.29}$$

We will show that $\mathfrak{A}^\tau$ is semisimple. Suppose now that $\pi(K) = 0$ for some $K \in \mathfrak{A}^\tau$. Then, since

$$(\lambda(K)|\pi(K_1)\lambda(A)) = (\lambda(A^\sharp)|\pi(K)^*\lambda(K_1)) = 0$$

for all $A \in \mathfrak{A}$ and $K_1 \in \mathfrak{A}^\tau$, it follows from (4.4.28) that $\lambda(K) = 0$ as well as $\lambda(K^\sharp) = 0$, which implies that $N(\mathfrak{A}^\tau) = \{O\}$, that is, $\mathfrak{A}^\tau$ is semisimple. Hence it follows from Theorem 4.2.8 that

$$\mathfrak{A}^T = \mathfrak{A}^{\tau\tau}. \tag{4.4.30}$$

Next we can prove that

$$\mathfrak{B}^\tau = \mathfrak{A}^\tau. \tag{4.4.31}$$

Indeed, since $[\nu(\mathfrak{B})] = [\nu(\mathfrak{A})]$, it follows that $\lambda(\mathfrak{B})$ is dense in $\lambda(\mathfrak{A})$, which yields by (4.4.28) that

$$[\pi(\mathfrak{A}^\tau)\lambda(\mathfrak{B})] = \mathcal{H}. \tag{4.4.32}$$

Take arbitrary $K \in \mathfrak{B}^\tau$ and $A \in \mathfrak{A}$. Then, because $[\nu(\mathfrak{B})] = [\nu(\mathfrak{A})]$, we can take a sequence $\{B_n\}$ in $\mathfrak{B}$ such that $\lim_{n\to\infty} \lambda(B_n) = \lambda(A)$ and $\lim_{n\to\infty} \lambda(B_n^\sharp) = \lambda(A^\sharp)$, and since $\mathfrak{B}^\tau = \mathfrak{B}^\rho$ by the semisimplicity of $\mathfrak{B}$, it follows that

$$\begin{aligned}(\lambda(A)|\lambda(K)) &= \lim_{n\to\infty} (\lambda(B_n)|\lambda(K)) \\ &= \lim_{n\to\infty} (\lambda(K^\sharp)|\lambda(B_n^\sharp)) \\ &= (\lambda(K^\sharp)|\lambda(A^\sharp)),\end{aligned} \tag{4.4.33}$$

which implies that

$$\begin{aligned}(\pi(A)\lambda(K)|\pi(K_1)\lambda(B)) &= (\pi(K_1)^*\lambda(K)|\pi(A)^*\lambda(B)) \\ &= (\lambda(K_1^\sharp K)|\lambda(A^\sharp B)) \\ &= (\lambda(B^\sharp A)|\lambda(K^\sharp K_1)) \\ &= (\pi(K)\lambda(A)|\pi(K_1)\lambda(B))\end{aligned}$$

for all $K_1 \in \mathfrak{A}^\tau$ and $B \in \mathfrak{B}$. Therefore we get, by (4.4.32),

$$\pi(A)\lambda(K) = \pi(K)\lambda(A). \tag{4.4.34}$$

Similarly, we can prove that

$$\pi(A)^*\lambda(K^\sharp) = \pi(K)^*\lambda(A^\sharp). \tag{4.4.35}$$

Furthermore, it follows from (4.4.28) that

$$\begin{aligned}\pi(K)\pi(A)\pi(K_1)\lambda(B) &= \pi(KK_1)\lambda(AB) \\ &= \pi(AB)\lambda(KK_1) \\ &= \pi(A)\pi(B)\pi(K)\lambda(K_1) \\ &= \pi(A)\pi(K)\pi(B)\lambda(K_1) \\ &= \pi(A)\pi(K)\pi(K_1)\lambda(B)\end{aligned}$$

for all $K_1 \in \mathfrak{A}^\tau$ and $B \in \mathfrak{B}$, which implies by (4.4.32) that

$$\pi(K)\pi(A) = \pi(A)\pi(K). \tag{4.4.36}$$

By (4.4.33)–(4.4.36), we have $K \in \mathfrak{A}^{\tau}$ and $\mathfrak{B}^{\tau} \subset \mathfrak{A}^{\tau}$. The converse inclusion: $\mathfrak{A}^{\tau} \subset \mathfrak{B}^{\tau}$ is trivial. Thus, (4.4.31) holds. By (4.4.30), (4.4.31) and Theorem 3.1.1, we have

$$\bar{\mathfrak{A}}[\tau_{\sigma s}^*] = \mathfrak{A}^T = \mathfrak{A}^{\tau\tau} = \mathfrak{B}^{\tau\tau} = \bar{\mathfrak{B}}[\tau_{\sigma s}^*].$$

This completes the proof.

The following is an immediate consequence of Theorem 4.4.14.

Corollary 4.4.16 *Let $\mathfrak{A}$ be a semisimple T^*-algebra on $\mathcal{H}$ and $\mathfrak{B}$ a $*$-subalgebra of $\mathfrak{A}$. Suppose that $\nu(\mathfrak{B})$ is dense in the Hilbert space $\mathcal{H} \oplus \mathcal{H}^*$. Then $\mathfrak{B}$ is σ-strongly $*$ dense in $\mathfrak{A}$.*

We next investigate the uniform density of a $*$-subalgebra of a T^*-algebra. Let $\mathfrak{L}$ be a vector space and r a norm on $\mathfrak{L}$. We denote by $\mathfrak{L}^*(r)$ the dual space of the normed space $\mathfrak{L}[r]$, and denote its unit ball by $u^*(r)$, that is,

$$u^*(r) = \{f \in \mathfrak{L}^*(r); \quad |f(x)| \leqq r(x) \quad \text{for all} \quad x \in \mathfrak{L}[r]\}.$$

Then we have the following

Lemma 4.4.17 *Let r_1 and r_2 be norms on $\mathfrak{L}$ and define the norm $r_1 \vee r_2$ on $\mathfrak{L}$ by*

$$(r_1 \vee r_2)(x) = \max(r_1(x), r_2(x)), \quad x \in \mathfrak{L}.$$

Then $u^*(r_1 \vee r_2)$ is the smallest convex set containing $u^*(r_1) \cup u^*(r_2)$, that is,

$$u^*(r_1 \vee r_2) = \{tf + (1-t)g; \quad 0 \leqq t \leqq 1, f \in u^*(r_1), \ g \in u^*(r_2)\}.$$

Proof Let W be the smallest convex set containing $u^*(r_1)$ and $u^*(r_2)$. Since the map:

$$(f, g, t) \in u^*(r_1) \times u^*(r_2) \times [0, 1] \longrightarrow tf + (1-t)g \in W$$

is weakly* continuous from the weakly compact set onto W, W is weakly compact. By the Mackey-Arens theorem [31], there exists a norm r on $\mathfrak{L}$ such that

$$u^*(r) = W \quad \text{and} \quad r(x) = \sup\{|f(x)|; \quad f \in W\} \quad \text{for all } x \in \mathfrak{L}.$$

Then we have

$$\begin{aligned} r(x) &= \sup\{|f(x)|; \quad f \in W\} \\ &= \sup\{|tf_1(x) + (1-t) f_2(x)|; \quad 0 \leqq t \leqq 1, f_1 \in u^*(r_1), \ f_2 \in u^*(r_2)\} \end{aligned}$$

$$= \max(r_1(x), r_2(x))$$
$$= (r_1 \vee r_2)(x)$$

for all $x \in \mathfrak{L}$, which completes the proof.

The next theorem is another main result of this subsection.

Theorem 4.4.18 *Let $\mathfrak{A}$ be a T^*-algebra on $\mathcal{H}$ and $\mathfrak{B}$ a $*$-subalgebra of $\mathfrak{A}$. Suppose that*

(i) $\pi(\mathfrak{B})$ is uniformly dense in $\pi(\mathfrak{A})$,
(ii) $N(\mathfrak{B})$ is strongly dense in $N(\mathfrak{A})$.
Then $\mathfrak{B}$ is uniformly dense in $\mathfrak{A}$.

Proof Since the uniform closures $\bar{\mathfrak{A}}[\tau_u]$ and $\bar{\mathfrak{B}}[\tau_u]$ of $\mathfrak{A}$ and $\mathfrak{B}$ satisfy assumptions (i) and (ii), we may assume that $\mathfrak{A}$ and $\mathfrak{B}$ are CT^*-algebras on $\mathcal{H}$. Then, it suffices to show that $\mathfrak{A} = \mathfrak{B}$. Suppose now that there exists an element X of $\mathfrak{A} \setminus \mathfrak{B}$. By the Hahn-Banach theorem [31] there exists a continuous linear functional on the Banach space $\bar{\mathfrak{A}}[\tau_u]$ such that

$$\|f\| = 1, \quad f(X) > 0 \quad \text{and} \quad f(\mathfrak{B}) = 0. \tag{4.4.37}$$

Put

$$r_\pi(A) = \|\pi(A)\|, \quad r_\lambda(A) = \|\lambda(A)\|, \quad r_{\lambda^*}(A) = \|\lambda(A^\sharp)\|$$

and

$$r(A) = \|A\| = \max\left(r_\pi(A), r_\lambda(A), r_{\lambda^*}(A)\right), \quad A \in \mathfrak{A}.$$

By Lemma 4.4.17, $u^*(r)$ is the smallest convex set containing $u^*(r_\pi)$, $u^*(r_\lambda)$ and $u^*(r_{\lambda^*})$; hence

$$f = t_1 f_1 + t_2 f_2 + t_3 f_3,$$

where $t_1 + t_2 + t_3 = 1$, $t_i \geqq 0$ (i=1,2,3) and $f_1 \in u^*(r_\pi)$, $f_2 \in u^*(r_\lambda)$ and $f_3 \in u^*(r_{\lambda^*})$. We here define a linear functional F on $\pi(\mathfrak{A})$ by

$$F(\pi(A)) = f_1(A), \quad A \in \mathfrak{A}.$$

Since $f_1 \in u^*(r_\pi)$, F can be extended to the continuous linear functional on the C^*-algebra $\overline{\pi(\mathfrak{A})}[\tau_u^\pi]$, which is denoted by the same F. By the GNS-construction for F there exist a Hilbert space $\mathcal{H}_F$, a $*$-homomorphism π_F of $\overline{\pi(\mathfrak{A})}[\tau_u^\pi]$ into $B(\mathcal{H}_F)$ and $x, y \in \mathcal{H}_F$ such that

$$f_1(A) = F(\pi(A)) = (\pi_F(\pi(A))x|y)$$

for all $A \in \mathfrak{A}$. We now put

$$\Phi_F(A) = \pi_F(\pi(A)), \quad A \in \mathfrak{A}.$$

Then Φ_F is a $*$-homomorphism of $\mathfrak{A}$ into $B(\mathcal{H}_F)$ satisfying

$$f_1(A) = (\Phi_F(A)x|y), \quad A \in \mathfrak{A}.$$

Furthermore, since $f_2 \in u^*(r_\lambda)$ and $f_3 \in u^*(r_{\lambda^*})$, it follows from the Riesz theorem that there exist elements ξ_1 and ξ_2 of $\mathcal{H}$ such that

$$\|\xi_1\| \leqq 1 \quad \text{and} \quad f_2(A) = (\lambda(A)|\xi_1),$$
$$\|\xi_2\| \leqq 1 \quad \text{and} \quad f_3(A) = (\xi_2|\lambda(A^\sharp))$$

for all $A \in \mathfrak{A}$. Hence

$$f(A) = t_1(\Phi_F(A)x|y) + t_2(\lambda(A)|\xi_1) + t_3(\xi_2|\lambda(A^\sharp)) \tag{4.4.38}$$

for all $A \in \mathfrak{A}$. We now write

$$\hat{\pi} = \pi \oplus \Phi_F,$$
$$\hat{\lambda} = \lambda \oplus 0,$$
$$\hat{A} = (\hat{\pi}(A), \hat{\lambda}(A), \hat{\lambda}(A^\sharp)^*), \quad A \in \mathfrak{A}.$$

Then it is easily proved that $\hat{\mathfrak{A}} := \{\hat{A}; \quad A \in \mathfrak{A}\}$ is a CT^*-algebra on $\mathcal{H} \oplus \mathcal{H}_F$. Furthermore, $\hat{\mathfrak{B}}$ and $\hat{\mathfrak{A}}$ satisfy assumptions (i) and (ii). Since $\pi(\hat{\mathfrak{B}})$ is also strongly dense in $\pi(\hat{\mathfrak{A}})$, it follows from Theorem 4.4.12 that $\hat{\mathfrak{B}}$ is strongly* dense in $\hat{\mathfrak{A}}$. Hence, for any $\varepsilon > 0$ there exists an element B of $\mathfrak{B}$ such that

$$\|\Phi_F(X)x - \Phi_F(B)x\| < \varepsilon,$$
$$\|\lambda(X) - \lambda(B)\| < \varepsilon,$$
$$\|\lambda(X^\sharp) - \lambda(B^\sharp)\| < \varepsilon,$$

which yields by (4.4.37) and (4.4.38) that

$$\begin{aligned}|f(X)| &= |f(X) - f(B)| \\ &= |t_1(\Phi_F(X)x - \Phi_F(B)x|y) + t_2(\lambda(X) - \lambda(B)|\xi_1) \\ &\quad + t_3(\xi_2|\lambda(X^\sharp) - \lambda(B^\sharp))|\end{aligned}$$

$$
\begin{aligned}
&\leqq \|\Phi_F(X)x - \Phi_F(B)x\| \|y\| + \|\lambda(X) - \lambda(B)\| \\
&\quad + \|\lambda(X^\sharp) - \lambda(B^\sharp)\| \\
&< (\|y\| + 2)\varepsilon.
\end{aligned}
$$

Hence $f(X) = 0$, which contradicts to (4.4.37). Thus, $\mathfrak{A} = \mathfrak{B}$. This completes the proof.

The following is an immediate consequence of Theorem 4.4.18.

Corollary 4.4.19 *Let $\mathfrak{B}$ be a $*$-subalgebra of a T^*-algebra $\mathfrak{A}$. Suppose that $\mathfrak{A}$ is semisimple and $\pi(\mathfrak{B})$ is uniformly dense in $\pi(\mathfrak{A})$. Then $\mathfrak{B}$ is uniformly dense in $\mathfrak{A}$.*

Chapter 5
Applications

This chapter is devoted to applications of the theory of observable algebras exhibited in Chaps. 2–4. In Sect. 5.1 standard T^*-algebras having a close relation to the so-called Tomita-Takesaki theory are discussed. Section 5.2 applies the theory of observable algebras to admissible (representable) i.p.s. forms on (locally convex) $*$-algebras. In Sect. 5.2.1 we define the notions of admissibility and of representability of i.p.s. forms φ on (locally convex) $*$-algebras, and construct the CT^*-algebra $\mathfrak{A}_\varphi$ from φ. In Sect. 5.2.2 we investigate admissibility and representability of i.p.s. forms on locally convex $*$-algebras. Section 5.2.3 gives necessary and sufficient conditions under which an admissible i.p.s. form φ on a $*$-algebra $\mathscr{A}$ is represented as $\varphi(a, b) = (\pi_\varphi(a)g|\pi_\varphi(b)g)$, $a, b \in \mathscr{A}$ for some $g \in \mathcal{H}_\varphi$. Section 5.2.4 shows that every admissible i.p.s. form φ on a $*$-algebra $\mathscr{A}$ is decomposed into the regular i.p.s. form φ_r and the singular i.p.s. form φ_s using the decomposition theorem of the CT^*-algebra $\mathfrak{A}_\varphi$ studied in Sect. 4.1. In Sect. 5.2.5 the regularity and the singularity of admissible i.p.s. forms are investigated applying the results of Sect. 4.2 to the CT^*-algebra $\mathfrak{A}_\varphi$. In Sect. 5.3 we show that every positive definite generalized function in a separable Lie group is representable, and apply the results in Sects. 4.1 and 4.2 to it. In Sect. 5.4 we discuss weights in C^*-algebras using the theory of CT^*-algebras.

5.1 Standard T^*-Algebras and the Tomita-Takesaki Theory

In this section we define the notion of standard T^*-algebras and show that it is closely related to the Tomita-Takesaki theory.

A. Inoue, *Tomita's Lectures on Observable Algebras in Hilbert Space*,
Lecture Notes in Mathematics 2285, https://doi.org/10.1007/978-3-030-68893-6_5

Definition 5.1.1 A T^*-algebra $\mathfrak{A}$ on $\mathcal{H}$ is called standard if it satisfies the following conditions:

(i) semisimple.
(ii) cyclic, that is, $\lambda(\mathfrak{A})$ is dense in $\mathcal{H}$.
(iii) separating, that is, $\lambda(\mathfrak{A}^\tau)$ is dense in $\mathcal{H}$.

Let $\mathfrak{A}$ be a standard T^*-algebra on $\mathcal{H}$. Then it is regular and its uniform closure $\bar{\mathfrak{A}}[\tau_u]$ is a standard CT^*-algebra on $\mathcal{H}$. We next give some definitions and terminology concerning left Hilbert algebras, right Hilbert algebras and modular Hilbert algebras. For detail, see [36].

Definition 5.1.2 A $*$-algebra $\mathfrak{A}$ with involution: $a \in \mathfrak{A} \to a^\sharp \in \mathfrak{A}$ (resp. $a \in \mathfrak{A} \to a^\flat \in \mathfrak{A}$) which is a dense subspace in a Hilbert space $\mathcal{H}$ is called a left (resp. right) Hilbert algebra in $\mathcal{H}$ if it satisfies the following conditions (i)–(iv):

(i) For any $a \in \mathfrak{A}$ the linear operator defined by $\pi_l(a)b = ab$ (resp. $\pi_r(a)b = ba$), $b \in \mathfrak{A}$ is bounded, and its extension on $\mathcal{H}$ is denoted by the same $\pi_l(a)$ (resp. $\pi_r(a)$).
(ii) $(ab|c) = (b|a^\sharp c)$ (resp. $(ab|c) = (a|cb^\flat)$) for each $a,\ b,\ c \in \mathfrak{A}$.
(iii) $\mathfrak{A}^2$ is dense in $\mathcal{H}$, where $\mathfrak{A}^2$ is a $*$-subalgebra spanned by $\{ab;\ a, b \in \mathfrak{A}\}$.
(iv) The involution: $a \in \mathfrak{A} \to a^\sharp \in \mathfrak{A}$ (resp. $a \in \mathfrak{A} \to a^\flat \in \mathfrak{A}$) is closable.

Definition 5.1.3 A left Hilbert algebra $\mathfrak{A}$ in $\mathcal{H}$ is called a modular Hilbert algebra if it admits a complex one-parameter group $\Delta(\alpha)$ of automorphisms of $\mathfrak{A}$ satisfying the following conditions (i)–(v):

(i) $(\Delta(\alpha)a)^\sharp = \Delta(-\bar{\alpha})a^\sharp$, $a \in \mathfrak{A}$, $\alpha \in \mathbb{C}$.
(ii) $(\Delta(\alpha)a|b) = (a|\,\Delta\,(\bar{\alpha})b)$, $a, b \in \mathfrak{A}$.
(iii) $(\Delta(1)a^\sharp|b^\sharp) = (b|a)$, $a, b \in \mathfrak{A}$.
(iv) $(\Delta(\alpha)a|b)$, $a, b \in \mathfrak{A}$, is an analytic function of α of $\mathbb{C}$.
(v) For every $t \in \mathbb{R}$, the set $(I + \Delta(t))\mathfrak{A}$ is dense in $\mathcal{H}$.

At present, a modular Hilbert algebra is called the *Tomita algebra*.

Let $\mathfrak{A}$ be a left Hilbert algebra in $\mathcal{H}$. Then, π_l is a nondegenerate $*$-representation of $\mathfrak{A}$ on $\mathcal{H}$, and the von Neumann algebra $\pi_l(\mathfrak{A})''$ generated by $\pi_l(\mathfrak{A})$ is called the *left von Neumann algebra* of $\mathfrak{A}$. Let $S_{\mathfrak{A}}$ denote the closure of the involution $a \in \mathfrak{A} \to a^\sharp \in \mathfrak{A}$ and let $S_{\mathfrak{A}} = J_{\mathfrak{A}}\Delta_{\mathfrak{A}}^{\frac{1}{2}}$ be the polar decomposition of the closed conjugate linear operator $S_{\mathfrak{A}}$. Then we have

(1) $J_{\mathfrak{A}}$ is a conjugate linear isometry on $\mathcal{H}$ satisfying $J_{\mathfrak{A}} = J_{\mathfrak{A}}^* = J_{\mathfrak{A}}^{-1}$;
(2) $S_{\mathfrak{A}}^2 x = x$, $x \in D(S_{\mathfrak{A}})$;
(3) $\Delta_{\mathfrak{A}} = S_{\mathfrak{A}}^* S_{\mathfrak{A}}$ is a non-singular positive self-adjoint operator in $\mathcal{H}$;
(4) $J_{\mathfrak{A}}\,\Delta_{\mathfrak{A}}\,J_{\mathfrak{A}} = \Delta_{\mathfrak{A}}^{-1}$, $S_{\mathfrak{A}} = J_{\mathfrak{A}}\Delta_{\mathfrak{A}}^{\frac{1}{2}} = \Delta_{\mathfrak{A}}^{-\frac{1}{2}}J_{\mathfrak{A}}$, $S_{\mathfrak{A}}^* = J_{\mathfrak{A}}\Delta_{\mathfrak{A}}^{-\frac{1}{2}} = \Delta_{\mathfrak{A}}^{\frac{1}{2}}J_{\mathfrak{A}}$.

The $J_{\mathfrak{A}}$ is called the *modular conjugation* of $\mathfrak{A}$ and $\Delta_{\mathfrak{A}}$ is called the *modular operator* of $\mathfrak{A}$.

We next define the commutant $\mathfrak{A}'$ of $\mathfrak{A}$. For any $b \in D(S_{\mathfrak{A}}^*)$ we define a linear operator $\pi_r(b)$ on $\mathfrak{A}$ by

$$\pi_r(b)a = \pi_l(a)b, \quad a \in \mathfrak{A}.$$

Then it is a closable operator in $\mathcal{H}$ satisfying $\pi_r(b)^* \supset \pi_r(S_{\mathfrak{A}}^* b)$. We denote its closure by the same $\pi_r(b)$, which is affiliated to $\pi_l(\mathfrak{A})'$. Here, the commutant $\mathfrak{A}'$ of $\mathfrak{A}$ is defined by

$$\mathfrak{A}' = \{b \in D(S_{\mathfrak{A}}^*); \quad \pi_r(b) \in B(\mathcal{H})\}.$$

In [36] Takesaki showed that $\mathfrak{A}'$ is dense in $\mathcal{H}$ and it is a right Hilbert algebra in $\mathcal{H}$ equipped with the following multiplication and the involution:

$$b_1 b_2 := \pi_r(b_2)b_1, \quad b_1, b_2 \in \mathfrak{A}';$$
$$b^\flat := S_{\mathfrak{A}}^* b, \quad b \in \mathfrak{A}',$$

satisfying $\pi_r(\mathfrak{A}')'' = \pi_l(\mathfrak{A})'$. In Propositions 5.1.4 and 5.1.5 later, we will show the above results using the T^*-algebra on $\mathcal{H}$ defined by

$$\tau_l(\mathfrak{A}) = \{\tau_l(a) := (\pi_l(a), a, (a^\sharp)^*); \quad a \in \mathfrak{A}\}.$$

Since $\mathfrak{A}'$ is dense in $\mathcal{H}$, the bicommutant $\mathfrak{A}'' := (\mathfrak{A}')'$ of $\mathfrak{A}$ is defined as follows: For any $a \in D(S_{\mathfrak{A}})$ we define a linear operator $\pi_l(a)$ by

$$\pi_l(a)b = \pi_r(b)a, \quad b \in \mathfrak{A}'.$$

Then $\pi_l(a)$ is a closable operator in $\mathcal{H}$ satisfying $\pi_l(a)^* \supset \pi_l(S_{\mathfrak{A}} a)$, and denote the closure of $\pi_l(a)$ by the same $\pi_l(a)$. We now put

$$\mathfrak{A}'' = \{a \in D(S_{\mathfrak{A}}); \quad \pi_l(a) \in B(\mathcal{H})\},$$

which is a full left Hilbert algebra in $\mathcal{H}$ equipped with the multiplication and the involution:

$$a_1 a_2 := \pi_l(a_1)a_2, \quad a_1, a_2 \in \mathfrak{A}'';$$
$$a^\sharp := S_{\mathfrak{A}} a, \quad a \in \mathfrak{A}'',$$

containing $\mathfrak{A}$. We here say that a left Hilbert algebra $\mathfrak{A}$ is called *full* if $\mathfrak{A} = \mathfrak{A}''$.

For standard T^*-algebras we have the following

Proposition 5.1.4 *Suppose that $\mathfrak{A}$ is a standard T^*-algebra on $\mathcal{H}$. Then, the subspace $\lambda(\mathfrak{A})$ in $\mathcal{H}$ is a left Hilbert algebra in $\mathcal{H}$ equipped with the following multiplication and the involution:*

$$\lambda(A_1)\lambda(A_2) = \lambda(A_1 A_2), \quad A_1, A_2 \in \mathfrak{A};$$
$$\lambda(A)^\sharp = \lambda(A^\sharp), \quad A \in \mathfrak{A},$$

satisfying

(i) $\lambda(\mathfrak{A})' = \lambda(\mathfrak{A}^\tau)$ *and*

$$\pi_r(\lambda(K)) = \pi(K), \quad \lambda(K)^\flat := S^*_{\lambda(\mathfrak{A})}\lambda(K) = \lambda(K^\sharp)$$

for all $K \in \mathfrak{A}^\tau$;

(ii) $\lambda(\mathfrak{A})'' = \lambda(\mathfrak{A}^{\tau\tau})$ *and*

$$\pi_l(\lambda(A)) = \pi(A), \quad \lambda(A)^\sharp := S_{\lambda(\mathfrak{A})}\lambda(A) = \lambda(A^\sharp)$$

for all $A \in \mathfrak{A}^{\tau\tau}$.

Proof By definition of the multiplication and the involution of $\lambda(\mathfrak{A})$ we have

$$\pi_l(\lambda(A)) = \pi(A)$$

for all $A \in \mathfrak{A}$ and

$$\begin{aligned}(\lambda(A)\lambda(B)|\lambda(C)) &= (\pi(A)\lambda(B)|\lambda(C)) \\ &= (\lambda(B)|\pi(A)^*\lambda(C)) \\ &= (\lambda(B)|\lambda(A)^\sharp\lambda(C))\end{aligned}$$

for all $A, B, C \in \mathfrak{A}$. Since $\mathfrak{A}$ is regular and cyclic, it follows that $[\lambda(\mathfrak{A}^2)]$ is dense in $\mathcal{H}$. Now, take an arbitrary sequence $\{A_n\}$ in $\mathfrak{A}$ such that $\lim_{n\to\infty} \lambda(A_n) = 0$ and $\lim_{n\to\infty} \lambda(A_n)^\sharp = x$ for some $x \in \mathcal{H}$. Then, since $\mathfrak{A}$ is regular, it follows from Theorem 4.2.7 that $\mathfrak{A}^\tau = \mathfrak{A}^\rho$, which implies that

$$\begin{aligned}(x|\lambda(K)) &= \lim_{n\to\infty} (\lambda(A_n^\sharp)|\lambda(K)) \\ &= \lim_{n\to\infty} (\lambda(K^\sharp)|\lambda(A_n)) \\ &= 0.\end{aligned}$$

Since $\lambda(\mathfrak{A}^\tau)$ is dense in $\mathcal{H}$, we have $x = 0$. Therefore, the involution: $\lambda(A) \to \lambda(A^\sharp)$ is closable. Thus $\lambda(\mathfrak{A})$ is a left Hilbert algebra in $\mathcal{H}$. We verify that $\lambda(\mathfrak{A})' =$

$\lambda(\mathfrak{A}^\tau)$. Take an arbitrary $K \in \mathfrak{A}^\tau (= \mathfrak{A}^\rho)$. Then we have

$$\begin{aligned}(S_{\lambda(\mathfrak{A})}\lambda(A)|\lambda(K)) &= (\lambda(A^\sharp)|\lambda(K)) \\ &= (\lambda(K^\sharp)|\lambda(A))\end{aligned}$$

for all $A \in \mathfrak{A}$, which yields that $\lambda(K) \in D(S^*_{\lambda(\mathfrak{A})})$ and $S^*_{\lambda(\mathfrak{A})}\lambda(K) = \lambda(K^\sharp)$, and that

$$\begin{aligned}\pi_r(\lambda(K))\lambda(A) &= \pi_l(\lambda(A))\lambda(K) \\ &= \pi(A)\lambda(K) \\ &= \pi(K)\lambda(A)\end{aligned}$$

for all $A \in \mathfrak{A}$. Hence, $\pi_r(\lambda(K)) = \pi(K) \in B(\mathcal{H})$. Thus $\lambda(K) \in \lambda(\mathfrak{A})'$ and $\lambda(\mathfrak{A}^\tau) \subset \lambda(\mathfrak{A})'$. Conversely, take an arbitrary $b \in \lambda(\mathfrak{A})'$. Then, it follows that

$$\begin{aligned}\pi_r(b)\pi(A)\lambda(B) &= \pi_r(b)\lambda(AB) \\ &= \pi(AB)b \\ &= \pi(A)\pi_r(b)\lambda(B)\end{aligned}$$

and

$$\begin{aligned}(\pi_r(b)^*\lambda(A^\sharp)|\lambda(B)) &= (\lambda(A^\sharp)|\pi(B)b) \\ &= (\lambda(B^\sharp A^\sharp)|b) \\ &= (S^*_{\lambda(\mathfrak{A})}b|\lambda(AB)) \\ &= (\pi(A)S^*_{\lambda(\mathfrak{A})}b|\lambda(B))\end{aligned}$$

for all $A, B \in \mathfrak{A}$, which implies that

$$K := (\pi_r(b), b, (S^*_{\lambda(\mathfrak{A})}b)^*) \in \mathfrak{A}^\tau \quad \text{and} \quad b = \lambda(K) \in \lambda(\mathfrak{A}^\tau).$$

Hence, $\lambda(\mathfrak{A})' \subset \lambda(\mathfrak{A}^\tau)$. Thus, (i) holds. We next show (ii). Take an arbitrary $A \in \mathfrak{A}^{\tau\tau}$. By (i) we have

$$\begin{aligned}(S^*_{\lambda(\mathfrak{A})}\lambda(K)|\lambda(A)) &= (\lambda(K^\sharp)|\lambda(A)) \\ &= (\lambda(A^\sharp)|\lambda(K))\end{aligned}$$

for all $K \in \mathfrak{A}^\tau (= \mathfrak{A}^\rho)$. Hence it follows that $\lambda(A) \in D(S^{**}_{\lambda(\mathfrak{A})}) = D(S_{\lambda(\mathfrak{A})})$ and $S_{\lambda(\mathfrak{A})}\lambda(A) = \lambda(A^\sharp)$, and that

$$\begin{aligned}\pi_l(\lambda(A))\lambda(K) &= \pi_r(\lambda(K))\lambda(A)\\ &= \pi(K)\lambda(A)\end{aligned}$$

for all $K \in \mathfrak{A}^\tau$, which yields that $\lambda(A) \in \lambda(\mathfrak{A})''$ and $\pi_l(\lambda(A)) = \pi(A)$ for all $A \in \mathfrak{A}^{\tau\tau}$. Thus, $\lambda(\mathfrak{A}^{\tau\tau}) \subset \lambda(\mathfrak{A})''$. Conversely, take an arbitrary $a \in \lambda(\mathfrak{A})''$. Then it follows, as in (i) that

$$A := (\pi_l(a), a, (S_{\lambda(\mathfrak{A})}a)^*) \in \mathfrak{A}^{\tau\tau} \quad \text{and} \quad a = \lambda(A) \in \lambda(\mathfrak{A}^{\tau\tau}),$$

so that $\lambda(\mathfrak{A})'' \subset \lambda(\mathfrak{A}^{\tau\tau})$. Thus, (ii) holds. This completes the proof.

We next show the converse of Proposition 5.1.4.

Proposition 5.1.5 *Suppose that $\mathfrak{A}$ is a left Hilbert algebra in $\mathcal{H}$. Then*

$$\tau_l(\mathfrak{A}) := \{\tau_l(a) := (\pi_l(a), a, (a^\sharp)^*); \quad a \in \mathfrak{A}\}$$

is a standard T^*-algebra on $\mathcal{H}$.

Proof It is clear that $\tau_l(\mathfrak{A})$ is a T^*-algebra on $\mathcal{H}$ satisfying

$$\pi(\tau_l(a)) = \pi_l(a), \quad \lambda(\tau_l(a)) = a \quad \text{and} \quad \lambda(\tau_l(a)^\sharp) = a^\sharp$$

for all $a \in \mathfrak{A}$. We here denote by $\mathfrak{A}_l$ the T^*-algebra $\tau_l(\mathfrak{A})$ for simplicity. We show that $\mathfrak{A}_l$ is regular. Since $\mathfrak{A}^2$ is dense in $\mathcal{H}$, we have $P_{\mathfrak{A}_l} = I$. Take an arbitrary $A = (0, \lambda(A), \lambda(A^\sharp)^*) \in N(\bar{\mathfrak{A}}_l[\tau_u])$. Then we can take a sequence $\{a_n\}$ in $\mathfrak{A}$ such that $\lim_{n\to\infty} \pi_l(a_n) = 0$, uniformly, $\lim_{n\to\infty} a_n = \lambda(A)$ and $\lim_{n\to\infty} a_n^\sharp = \lambda(A^\sharp)$. By the calculation:

$$\begin{aligned}(\pi_l(a)^*\lambda(A)|S^*_{\mathfrak{A}}b) &= (\lambda(A)|\pi_r(S^*_{\mathfrak{A}}b)a)\\ &= \lim_{n\to\infty}(a_n|\pi_r(S^*_{\mathfrak{A}}b)a)\\ &= \lim_{n\to\infty}(\pi_l(a_n)b|a)\\ &= 0\end{aligned}$$

for all $a \in \mathfrak{A}$ and $b \in D(S^*_{\mathfrak{A}})$, we have $\pi_l(a)^*\lambda(A) = 0$ for all $a \in \mathfrak{A}$, which implies that $\lambda(A) = 0$ since $\mathfrak{A}^2$ is dense in $\mathcal{H}$. Similarly, $\lambda(A^\sharp) = 0$. Thus, $N(\bar{\mathfrak{A}}_l[\tau_u]) = \{O\}$, so $P_{\mathfrak{A}_l^\tau} = I$. This means that $\mathfrak{A}_l$ is regular. Finally we show that $\lambda(\mathfrak{A}_l^\tau)$ is dense in $\mathcal{H}$. Since

$$(I - C_{\mathfrak{A}_l^\tau})^\sim = (I - C_{\mathfrak{A}_l^\tau}, 0, 0) \in \mathfrak{A}^{\tau\tau},$$

where $C_{\mathfrak{A}_l^\tau} := \text{Proj}\,[\lambda(\mathfrak{A}_l^\tau)]$, and $\mathfrak{A}_l$ is regular, it follows from Theorem 3.1.1 that $(I - C_{\mathfrak{A}_l^\tau})^\sim \in \bar{\mathfrak{A}}_l[\tau_{\sigma s}^*]$. Hence we can take a sequence $\{a_n\}$ in $\mathfrak{A}$ such that $\pi_l(a_n) \to I - C_{\mathfrak{A}_l^\tau}$, σ-strongly *, $\lim_{n\to\infty} a_n = \lim_{n\to\infty} a_n^\sharp = 0$, so that for any $b \in D(S_{\mathfrak{A}}^*)$

$$\lim_{n\to\infty} \pi_r(b)a_n = \lim_{n\to\infty} \pi_l(a_n)b = (I - C_{\mathfrak{A}_l^\tau})b.$$

Since $\pi_r(b)$ is a closed operator in $\mathcal{H}$, it follows that $(I - C_{\mathfrak{A}_l^\tau})b = 0$ for all $b \in D(S_{\mathfrak{A}}^*)$. Therefore $C_{\mathfrak{A}_l^\tau} = I$, that is, $\lambda(\mathfrak{A}_l^\tau)$ is dense in $\mathcal{H}$. This completes the proof.

Notes Let $\mathfrak{A}$ be a left Hilbert algebra in $\mathcal{H}$. In Proposition 5.1.5, we have shown without assumption that $\mathfrak{A}'$ is dense in $\mathcal{H}$ that $\tau_l(\mathfrak{A})$ is a standard T^*-algebra on $\mathcal{H}$. This fact leads by Proposition 5.1.4 that $\mathfrak{A}' = \lambda(\tau_l(\mathfrak{A})^\tau)$, and that $\mathfrak{A}'$ and $(\mathfrak{A}')^2$ are dense in $\mathcal{H}$. In [39] Tomita defined the notion of *generalized Hilbert algebras* as follows: such an algebra is a $*$-algebra $\mathfrak{A}$ with involution: $a \in \mathfrak{A} \to a^\sharp \in \mathfrak{A}$ which is dense in a Hilbert space $\mathcal{H}$ satisfying the conditions (i)–(iii) in Definition 5.1.3 and the following condition (iv)$'$ instead of (iv) in Definition 5.1.3:

(iv)$'$ $\lambda(\tau_l(\mathfrak{A})^\tau) = \mathfrak{A}'$ is dense in $\mathcal{H}$,

which is equivalent to the notion of a left Hilbert algebra in Definition 5.1.2. And he constructed a modular Hilbert algebra $\mathfrak{B}$ in $\mathcal{H}$ which is equivalent to $\mathfrak{A}$, that is, $\mathfrak{B}'' = \mathfrak{A}''$, and using this, he obtained the following results so-called the Tomita theorem:

(1) $J_{\mathfrak{A}}\pi_l(\mathfrak{A})''J_{\mathfrak{A}} = \pi_r(\mathfrak{A})'$ and $\Delta_{\mathfrak{A}}^{it}\pi_l(\mathfrak{A})''\Delta_{\mathfrak{A}}^{-it} = \pi_l(\mathfrak{A})''$ for all $t \in \mathbb{R}$.
(2) $J_{\mathfrak{A}}\mathfrak{A}'' = \mathfrak{A}'$ and $\pi_r(J_{\mathfrak{A}}a) = J_{\mathfrak{A}}\pi_l(a)J_{\mathfrak{A}}$ for all $a \in \mathfrak{A}''$;

$$\Delta_{\mathfrak{A}}^{it}\mathfrak{A}'' = \mathfrak{A}'' \quad \text{and} \quad \pi_l(\Delta_{\mathfrak{A}}^{it}a) = \Delta_{\mathfrak{A}}^{it}\pi_l(a)\,\Delta_{\mathfrak{A}}^{-it} \quad \text{for all} \quad a \in \mathfrak{A}''.$$

Similar statements hold for $\mathfrak{A}'$.

In [36] Takesaki defined the notion of left Hilbert algebras $\mathfrak{A}$ in Definition 5.1.2 and constructed a modular Hilbert algebra which is equivalent to $\mathfrak{A}$ investigating the modular operator $\Delta_{\mathfrak{A}}$ and the modular conjugation $J_{\mathfrak{A}}$ in detail. Moreover, he proved the above Tomita theorem, and applied this theory to quantum physics. His lecture note [36] influenced greatly the Connes study of type III von Neumann algebras [7]. At present, this theory is called the Tomita-Takesaki theory. In [44], Van Daele gave a fairly easy proof of the Tomita theorem without the use of modular Hilbert algebras.

5.2 Admissible Invariant Positive Invariant Sesquilinear Forms on a ∗-Algebra

In this section we discuss admissible invariant positive sesquilinear forms and admissible positive linear functionals on $*$-algebras using the theory of CT^*-

algebras. Let $\mathscr{A}$ be a $*$-algebra. A form φ is called a *sesquilinear form* on $\mathscr{A} \times \mathscr{A}$ if

(i) $\varphi(\alpha a + \beta b, c) = \alpha\varphi(a, c) + \beta\varphi(b, c)$,
(ii) $\varphi(a, \alpha b + \beta c) = \bar{\alpha}\varphi(a, b) + \bar{\beta}\varphi(a, c)$
for all $a,\ b,\ c \in \mathscr{A}$ and $\alpha,\ \beta \in \mathbb{C}$. A sesquilinear form φ on $\mathscr{A} \times \mathscr{A}$ is called *positive* if
(iii) $\varphi(a, a) \geqq 0$ for all $a \in \mathscr{A}$,
and *invariant* if
(iv) $\varphi(ab, c) = \varphi(b, a^*c)$ for all $a,\ b,\ c \in \mathscr{A}$.

Henceforth, we abbreviate an invariant positive sesquilinear form on $\mathscr{A} \times \mathscr{A}$ as an *i.p.s. form on* $\mathscr{A}$.

5.2.1 T^*-Algebras Generated by Admissible i.p.s. Forms

Let φ be an i.p.s. form on a $*$-algebra $\mathscr{A}$. Then it is easily proved that

$$\varphi(a, b) = \overline{\varphi(b, a)} \quad \text{(hermitian)} \tag{5.2.1}$$

and

$$|\varphi(a, b)|^2 \leqq \varphi(a, a)\varphi(b, b) \quad \text{(Schwarz inequality)} \tag{5.2.2}$$

for all $a,\ b \in \mathscr{A}$. By (5.2.1) and (5.2.2) it is verified that $N_\varphi := \{a \in \mathscr{A};\ \varphi(a, a) = 0\}$ is a left ideal of $\mathscr{A}$, and denote by $\lambda_\varphi(a)$ the coset of $\mathscr{A}/_{N_\varphi}$ which contains $a \in \mathscr{A}$. Then we can define an inner product $(\cdot|\cdot)$ on the vector space $\lambda_\varphi(\mathscr{A}) = \mathscr{A}/_{N_\varphi}$ by

$$(\lambda_\varphi(a)|\lambda_\varphi(b)) = \varphi(a, b), \quad a, b \in \mathscr{A}.$$

Denote by $\mathcal{H}_\varphi$ the Hilbert space which is the completion of the pre-Hilbert space $\lambda_\varphi(\mathscr{A})$. Furthermore, for any $a \in \mathscr{A}$ we define a linear operator $\pi_\varphi(a)$ on $\lambda_\varphi(\mathscr{A})$ by

$$\pi_\varphi(a)\lambda_\varphi(b) = \lambda_\varphi(ab), \quad b \in \mathscr{A}.$$

In case that $\pi_\varphi(a)$ is bounded, we denote the extension of $\pi_\varphi(a)$ on $\mathcal{H}_\varphi$ by the same $\pi_\varphi(a)$. If $\pi_\varphi(a) \in B(\mathcal{H}_\varphi)$ for all $a \in \mathscr{A}$, then it is easily shown that π_φ is a $*$-homomorphism of $\mathscr{A}$ into $B(\mathcal{H}_\varphi)$, which is called a *$*$-representation of $\mathscr{A}$ on $\mathcal{H}_\varphi$*. In general, $\pi_\varphi(a)$ is not necessarily bounded for all $a \in \mathscr{A}$, and then we need to consider an *unbounded $*$-representation* of $\mathscr{A}$ (refer to [17, 28, 34]) but in this case we treat only the bounded case. So, we define the following notions:

Definition 5.2.1 An i.p.s. form φ on a $*$-algebra $\mathscr{A}$ is called admissible if for any $a \in \mathscr{A}$ there exists a positive constant γ_a depending a such that

$$\varphi(ab, ab) \leqq \gamma_a \varphi(b, b)$$

for all $b \in \mathscr{A}$, equivalently $\pi_\varphi(a) \in B(\mathcal{H}_\varphi)$ for all $a \in \mathscr{A}$. In case that $\mathscr{A}$ is a locally convex $*$-algebra, an admissible i.p.s. form φ on $\mathscr{A}$ is also called algebraically representable, and φ is called representable if for any $a \in \mathscr{A}$ there exists a continuous seminorm p on $\mathscr{A}$ such that

$$\varphi(ab, ab) \leqq p(a)^2 \varphi(b, b)$$

for all $b \in \mathscr{A}$, equivalently π_φ is a continuous $*$-representation of $\mathscr{A}$ on $\mathcal{H}_\varphi$.

For an admissible i.p.s. form φ on $\mathscr{A}$ we can construct the triplet $(\pi_\varphi, \lambda_\varphi, \mathcal{H}_\varphi)$ of the Hilbert space $\mathcal{H}_\varphi$, the $*$-representation π_φ of $\mathscr{A}$ on $\mathcal{H}_\varphi$ and the vector representation λ_φ of $\mathscr{A}$ to $\mathcal{H}_\varphi$, which is called the *GNS-construction for* φ. Furthermore, we can define a T^*-algebra $\tau_\varphi(\mathscr{A})$ on $\mathcal{H}_\varphi$ by

$$\tau_\varphi(\mathscr{A}) = \{\tau_\varphi(a) := (\pi_\varphi(a), \lambda_\varphi(a), \lambda_\varphi(a^*)^*); \quad a \in \mathscr{A}\},$$

and denote by $\mathfrak{A}_\varphi$ the uniform closure $\overline{\tau_\varphi(\mathscr{A})}[\tau_u]$ of the T^*-algebra $\tau_\varphi(\mathscr{A})$. Then, $\tau_\varphi(\mathscr{A})$ and $\mathfrak{A}_\varphi$ are called *the* T^**-algebra and the* CT^**-algebra generated by* φ, respectively.

A linear functional f on $\mathscr{A}$ is called *positive* if $f(a^*a) \geqq 0$ for all $a \in \mathscr{A}$. For a positive linear functional f on $\mathscr{A}$ we have

$$f(ab) = \overline{f(b^*a^*)} \quad \text{for all} \quad a, b \in \mathscr{A},$$

however, in case that $\mathscr{A}$ doesn't have identity, f is not necessarily *hermitian*, namely

$$f(a^*) = \overline{f(a)} \quad \text{for all} \quad a \in \mathscr{A}.$$

Let f be a positive linear functional on $\mathscr{A}$. Then we can define an i.p.s. form f^0 on $\mathscr{A}$ by

$$f^0(a, b) = f(b^*a), \quad a, b \in \mathscr{A},$$

which is called the *i.p.s. form induced by* f. Conversely, an i.p.s. form on $\mathscr{A}$ is not necessarily induced by a positive linear functional on $\mathscr{A}$. For example, the i.p.s. form φ on the CT^*-algebra $T^*(\mathcal{H})$ defined by

$$\varphi(A, B) = (\lambda(A)|\lambda(B)), \quad A, B \in T^*(\mathcal{H})$$

is not induced by any positive linear functionals on $T^*(\mathcal{H})$. In Sect. 5.2.3 we will investigate when an admissible i.p.s. form is induced by a positive linear functional. We define the notions of the addmissibility and the representability of positive linear functionals as follows:

Definition 5.2.2 A positive linear functional f on $\mathscr{A}$ is called admissible if f^0 is admissible. In case that $\mathscr{A}$ is a locally convex $*$-algebra, a positive linear functional f is called algebraically representable (resp. representable) if f^0 is algebraically representable (resp. representable).

Let f be a positive linear functional on $\mathscr{A}$. We denote the GNS-construction $(\pi_{f^0}, \lambda_{f^0}, \mathcal{H}_{f^0})$ for f^0 by $(\pi_f, \lambda_f, \mathcal{H}_f)$, and call it the *GNS-construction for* f.

5.2.2 *Admissibility and Representability of i.p.s. Forms*

In this subsection we discuss the admissibility and the representability of i.p.s. forms on (locally convex) $*$-algebras. We first consider some sufficient and necessary conditions under which a $*$-algebra has an admissible i.p.s. form. For this problem we obtain the following result: a $*$-algebra $\mathscr{A}$ has an admissible i.p.s. form if and only if $\mathscr{A}$ has an admissible positive linear functionals if and only if there exists a $*$-representation of $\mathscr{A}$ on a Hilbert space if and only if there exists a C^*-seminorm on $\mathscr{A}$. We next consider the case where a $*$-algebra is locally convex. A locally convex $*$-algebra $\mathscr{A}$ is called *algebraically representable* if every i.p.s. form on $\mathscr{A}$ is algebraically representable, and $\mathscr{A}$ is called *representable* if every continuous i.p.s. form on $\mathscr{A}$ is representable. We here investigate which locally convex $*$-algebras are (algebraically) representable.

Let $\mathscr{A}$ be a locally convex $*$-algebra. An element a of $\mathscr{A}$ is called *(Allan) bounded* if there exists a positive constant λ such that $\{(\lambda^{-1}a)^n; \quad n \in \mathbb{N}\}$ is a bounded subset of $\mathscr{A}$. The set $\mathscr{A}_0$ of all Allan bounded elements of $\mathscr{A}$ is called the bounded part of $\mathscr{A}$. We denote by $\mathscr{A}[\mathscr{A}_0^h]$ the $*$-subalgebra of $\mathscr{A}$ generated by $\mathscr{A}_0^h := \{a \in \mathscr{A}_0; \quad a = a^*\}$. If $\mathscr{A}_0$ is a $*$-subalgebra of $\mathscr{A}$ (in particular, if $\mathscr{A}$ is commutative), then $\mathscr{A}_0 = \mathscr{A}[\mathscr{A}_0^h]$. However, $\mathscr{A}_0$ is not even a subspace in general. We denote by $\mathscr{B}$ the collection of all bounded subsets $\boldsymbol{B}$ of $\mathscr{A}$ such that $\boldsymbol{B}$ is closed and absolutely convex, and $\boldsymbol{B}^2 \subset \boldsymbol{B}$. For any $\boldsymbol{B} \in \mathscr{B}$ we put

$$\mathscr{A}[\boldsymbol{B}] = \{\lambda x; \quad \lambda \in \mathbb{C}, \ x \in \boldsymbol{B}\},$$

$$\|x\|_{\boldsymbol{B}} = \inf\{t > 0; \quad x \in t\boldsymbol{B}\}, \quad x \in \mathscr{A}[\boldsymbol{B}].$$

Then $\mathscr{A}[\boldsymbol{B}]$ is a normed algebra under the norm $\|\cdot\|_{\boldsymbol{B}}$ corresponding to the Minkowski functional with respect to B. If the normed algebra $\mathscr{A}[\boldsymbol{B}]$ is complete for every $\boldsymbol{B} \in \mathscr{B}$, then $\mathscr{A}$ is called *pseudo-complete*. If $\mathscr{A}$ is sequentially complete, then it is pseudo-complete, but the converse does not hold in general. We write

$$\mathscr{B}^* = \{B \in \mathscr{B};\ B^* = B\}.$$

For any $B \in \mathscr{B}$, $A[\boldsymbol{B}]$ is a normed $*$-algebra with norm $\|\cdot\|_B$, and if $\mathscr{A}$ is pseudo-complete, then it is a Banach $*$-algebra. The radius $\beta(a)$ of boundedness of $a \in \mathscr{A}$ is defined by

$$\beta(a) = \inf\{t > 0;\ \{(t^{-1}a)^n;\quad n \in \mathbb{N}\} \text{ is bounded}\},$$

where $\inf \emptyset = \infty$, and it satisfies the following equalities:

$$\begin{aligned} \beta(a) &= \sup\{\overline{\lim_{n\to\infty}} |f(a^n)|^{\frac{1}{n}};\quad f \in \mathscr{A}'\} \\ &= \sup\{\overline{\lim_{n\to\infty}}\, p(a^n)^{\frac{1}{n}};\quad p \in \mathscr{P}\}, \end{aligned} \tag{5.2.3}$$

where $\mathscr{A}'$ is the dual space of $\mathscr{A}$ and $\mathscr{P}$ is the family of seminorms which define the topology of $\mathscr{A}$. For more detail, refer to [1, 2, 10].

An element a of $\mathscr{A}$ is *quasi-regular* (resp. *quasi-invertible*) if $(e-a)$ has the inverse belonging to the unitization $\mathscr{A}_e$ of $\mathscr{A}$ (resp. belonging to $(\mathscr{A}_0)_e$). Let $\mathscr{A}^{qr}$ (resp. $\mathscr{A}^{qi}$) be the set of all quasi-regular (resp. quasi-invertible) elements of $\mathscr{A}$. The quasi-inverse of $a \in \mathscr{A}^{qr}$ is denoted by a^q, that is, $(e - a^q)(e - a) = (e - a)(e - a^q) = e$.

By ([5, Theorem 4.2]) we have the following

Proposition 5.2.3 *Let $\mathscr{A}$ be a pseudo-complete locally convex $*$-algebra. Consider the following statements:*

- *(i) $\mathscr{A}$ has a continuous quasi-inverse, that is, there exists a neighborhood U of 0 such that $U \subset \mathscr{A}^{qr}$ and the quasi-inversion $a \to a^q$ is continuous at 0.*
- *(ii) $\mathscr{A}^{qi}$ is open.*
- *(iii) There exists a continuous seminorm p on $\mathscr{A}$ such that $\beta(a) \leqq p(a)$ for all $a \in \mathscr{A}$.*
- *(iv) $\mathscr{A} = \mathscr{A}_0$.*

Then the following implications hold:

$$(i) \Longrightarrow \begin{array}{c}(ii)\\ \Updownarrow \\ (iii)\end{array} \Longrightarrow (iv).$$

If the multiplication of $\mathscr{A}$ is jointly continuous (e.g. $\mathscr{A}$ is a Fréchet $*$-algebra), then the above statements (i), (ii) and (iii) are equivalent.

We first deal with the algebraical representability of positive linear functionals and i.p.s.forms on a locally convex $*$-algebra $\mathscr{A}$ with $\mathscr{A} = \mathscr{A}_0$. For that, we define the following notion: An i.p.s. form φ on $\mathscr{A}$ is *weakly continuous* if for any $x \in \mathscr{A}$ the positive linear functional φ_x on $\mathscr{A}$ defined by

$$\varphi_x(a) = \varphi(ax, x), \quad a \in \mathscr{A} \tag{5.2.4}$$

is continuous. The weak continuity of φ derives that for any $x, y \in \mathscr{A}$ the linear functional $\varphi_{x,y}$ on $\mathscr{A}$ defined by

$$\varphi_{x,y}(a) = \varphi(ax, y), \quad a \in \mathscr{A}$$

is continuous because of the equality:

$$\begin{aligned}\varphi(ax, y) = \frac{1}{4}\{&\varphi(a(x+y), x+y) - \varphi(a(x-y), x-y) \\ &+ i\varphi(a(x+iy), x+iy) - i\varphi(a(x-iy), x-iy)\}.\end{aligned}$$

From this reason, an i.p.s. form satisfying (5.2.4) is called weakly continuous. We now have the following

Theorem 5.2.4 *Let $\mathscr{A}$ be a locally convex $*$-algebra with $\mathscr{A} = \mathscr{A}_0$. Then the following statements hold;*

(1) Suppose that $\mathscr{A}$ is pseudo-complete. Then every positive linear functional f on $\mathscr{A}$ is algebraically representable and satisfies the inequality:

$$f(x^*hx) \leqq \beta(h) f(x^*x) \tag{5.2.5}$$

for all $h \in \mathscr{A}_h := \{x \in \mathscr{A};\ x^* = x\}$ and $x \in \mathscr{A}$.

(2) Every continuous positive linear functional f is algebraically representable and satisfies the inequality (5.2.5).

(3) Every weakly continuous i.p.s. form φ on $\mathscr{A}$ is algebraically representable and satisfies the inequality:

$$|\varphi(hx, x)| \leqq \beta(h)\varphi(x, x)$$

for all $h \in \mathscr{A}_h$ and $x \in \mathscr{A}$.

Proof

(1) For any $h \in \mathscr{A}_h$ there exists an element $b \in \mathscr{A}_h$ such that

$$x^*h^2x + \beta(h)^2(x - bx)^*(x - bx) = \beta(h)^2 x^*x \tag{5.2.6}$$

for all $x \in \mathscr{A}$. Indeed, take an arbitrary $\varepsilon > 0$ and put

$$x_0 = \frac{1}{\beta(h)^2 + \varepsilon} h^2.$$

Since $\mathscr{A} = \mathscr{A}_0$, we get $x_0 \in \mathscr{A}_0^h$, so that by ([1, Prop. 2.4 and Cor. 2.17]), we can take an element $\boldsymbol{B}$ of $\mathscr{B}$ such that $x_0 \in \mathscr{A}[\boldsymbol{B}]$ and $\|x_0\|_{\boldsymbol{B}} < 1$. By the pseudo-completeness of $\mathscr{A}$, $(\mathscr{A}[\boldsymbol{B}], \|\cdot\|_{\boldsymbol{B}})$ is a Banach $*$-algebra. Hence, the series $\sum_{n=1}^{\infty} \binom{\frac{1}{2}}{n} (-x_0)^n$ in $\mathscr{A}[\boldsymbol{B}]$ converges to an element b of $\mathscr{A}_h$ and

$$2b - b^2 = x_0 = \frac{1}{\beta(h)^2 + \varepsilon} h^2.$$

Therefore we can prove that

$$x^* h^2 x + (\beta(h)^2 + \varepsilon)(x - bx)^*(x - bx) = (\beta(h)^2 + \varepsilon) x^* x$$

for all $x \in \mathscr{A}$, which implies (5.2.6) as $\varepsilon \to 0$. Let f be any positive linear functional on $\mathscr{A}$. By (5.2.6) we have

$$f(x^* h^2 x) \leqq \beta(h)^2 f(x^* x) \tag{5.2.7}$$

for all $x \in \mathscr{A}$. Hence we get that $\pi_f(h) \in B(\mathcal{H}_f)$ and $\|\pi_f(h)\| \leqq \beta(h)$, which yields that $\pi_f(a) \in B(\mathcal{H}_f)$ for all $a \in \mathscr{A}$, equivalently, f is algebraically representable.

(2) Let f be a continuous positive linear functional on $\mathscr{A}$. For an arbitrary $x \in \mathscr{A}$, define a continuous positive linear functional f_x on $\mathscr{A}$ by

$$f_x(a) = f(x^* a x), \quad a \in \mathscr{A}.$$

Using the Schwarz inequality we can prove that

$$|f_x(h)|^{2^n} \leqq f(x^* x)^{2^n - 1} f_x(h^{2^n})$$

for all $h \in \mathscr{A}_h$ and $n \in \mathbb{N}$, and by the continuity of f and (5.2.3) that

$$f(x^* h x) \leqq f(x^* x) \beta(h)$$

for all $h \in \mathscr{A}_h$ and $x \in \mathscr{A}$. This implies that $\pi_f(h) \in B(\mathcal{H}_f)$ and $\|\pi_f(h)\| \leqq \beta(h)$ for all $h \in \mathscr{A}_h$, so that $\pi_f(a) \in B(\mathcal{H}_f)$ for all $a \in \mathscr{A}$.

(3) Let φ be a weakly continuous i.p.s. form on $\mathscr{A}$. Using the Schwarz inequality of the positive linear functional φ_x on $\mathscr{A}$ we get that

$$|\varphi_x(h)|^{2^n} \leqq \varphi(x, x)^{2^n - 1} \varphi_x(h^{2^n})$$

for all $h \in \mathscr{A}_h$ and $n \in \mathbb{N}$ and by the continuity of φ_x and (5.2.3)

$$|\varphi_x(h)| \leqq \varphi(x, x) \beta(h).$$

Hence it follows that $\pi_\varphi(h) \in B(\mathcal{H}_\varphi)$ and $\|\pi_\varphi(h)\| \leqq \beta(h)$ for all $h \in \mathscr{A}_h$, which implies that φ is algebraically representable. This completes the proof.

For the representability of positive linear functionals and i.p.s. forms we have the following

Theorem 5.2.5 *Let $\mathscr{A}$ be a locally convex $*$-algebra. Suppose that there exists a continuous seminorm p on $\mathscr{A}$ such that $\beta(a) \leqq p(a)$ for all $a \in \mathscr{A}$. Then the following statements hold:*

(1) Suppose that $\mathscr{A}$ pseudo-complete. Then every positive linear functional on $\mathscr{A}$ is representable.
(2) Every continuous positive linear functional on $\mathscr{A}$ is representable.
(3) Every weakly continuous i.p.s. form on $\mathscr{A}$ is representable.

Proof By the continuity of the involution on $\mathscr{A}$ we may assume that the continuous seminorm p on $\mathscr{A}$ satisfies that $p(a^*) = p(a)$ for all $a \in \mathscr{A}$.
(1) By Proposition 5.2.3 we get $\mathscr{A} = \mathscr{A}_0$, so that Theorem 5.2.4 shows that every positive linear functional f on $\mathscr{A}$ is algebraically representable and $\|\pi_f(h)\| \leqq \beta(h)$ for all $h \in \mathscr{A}_h$. Hence we have

$$\begin{aligned}\|\pi_f(a)\| &\leqq \left\|\pi_f\left(\frac{a+a^*}{2}\right)\right\| + \left\|\pi_f\left(\frac{a-a^*}{2i}\right)\right\| \\ &\leqq \beta\left(\frac{a+a^*}{2}\right) + \beta\left(\frac{a-a^*}{2i}\right) \\ &\leqq p\left(\frac{a+a^*}{2}\right) + p\left(\frac{a-a^*}{2i}\right) \\ &\leqq 2p(a)\end{aligned}$$

for all $a \in \mathscr{A}$, which implies that f is representable. Similarly, we can show (2) and (3). This completes the proof.

In case of Banach $*$-algebras we have the following

Corollary 5.2.6 *Let $\mathscr{A}$ be a Banach $*$-algebra. Then every positive linear functional f on $\mathscr{A}$ and every i.p.s. form φ on $\mathscr{A}$ are both representable, and*

$$\|\pi_f(a)\| \leqq \|a\| \quad \textit{and} \quad \|\pi_\varphi(a)\| \leqq \|a\|$$

for all $a \in \mathscr{A}$.

Proof Since $\mathscr{A}$ is a Banach $*$-algebra, we have

$$\beta(a) \leqq \|a\| \quad \text{for all} \quad a \in \mathscr{A}.$$

Hence it follows from Theorem 5.2.5 that every positive linear functional f on $\mathscr{A}$ is representable and $\|\pi_f(a)\| \leqq \|a\|$ for all $a \in \mathscr{A}$. Let φ be an i.p.s. form on $\mathscr{A}$. Take an arbitrary $x \in \mathscr{A}$, and put

$$\tilde{\varphi_x}(a + \alpha e) = \varphi(ax + \alpha x, x)$$

for $a \in \mathscr{A}$ and $\alpha \in \mathbb{C}$. Then $\tilde{\varphi_x}$ is a positive linear functional on the Banach $*$-algebra $\mathscr{A}_e$ (the unitization of $\mathscr{A}$). Indeed, it is clear that $\tilde{\varphi_x}$ is a linear functional on $\mathscr{A}_e$ and

$$\begin{aligned}\tilde{\varphi_x}\left((a + \alpha e)^*(a + \alpha e)\right) &\geqq \left(\varphi(ax, ax)^{\frac{1}{2}} - |\alpha|\varphi(x, x)^{\frac{1}{2}}\right)^2 \\ &\geqq 0\end{aligned}$$

for all $a \in \mathscr{A}$ and $\alpha \in \mathbb{C}$. Hence it follows from [20] that $\tilde{\varphi_x}$ is continuous, which implies that φ is weakly continuous, so that by Theorem 5.2.5, φ is representable and $\|\pi_\varphi(a)\| \leqq \|a\|$ for all $a \in \mathscr{A}$. This completes the proof.

We consider the case of a locally convex $*$-algebra where $\mathscr{A} \neq \mathscr{A}_0$. A *locally m-convex $*$-algebra* is a locally convex $*$-algebra whose topology is determined by a family $\{p_\lambda\}_{\lambda\in\Lambda}$ of $*$-preserving and submultiplicative seminorms, that is, $p_\lambda(a^*) = p_\lambda(a)$ and $p_\lambda(ab) \leqq p_\lambda(a)p_\lambda(b)$ for all $\lambda \in \Lambda$ and $a, b \in \mathscr{A}$ [19].

Proposition 5.2.7 *Let $\mathscr{A}$ be a locally m-convex $*$-algebra. Then the following statements hold:*

(1) Every continuous positive linear functional on $\mathscr{A}$ is representable.
(2) Every weakly continuous i.p.s. form on $\mathscr{A}$ is representable.

Proof Let $\{p_\lambda\}_{\lambda\in\Lambda}$ be a family of $*$-preserving and submultiplicative seminorms which determines the topology of $\mathscr{A}$.

(1) Let f be any continuous positive linear functional on $\mathscr{A}$. Then there exist a positive constant $\gamma > 0$ and $\lambda \in \Lambda$ such that

$$|f(a)| \leqq \gamma p_\lambda(a) \tag{5.2.8}$$

for all $a \in \mathscr{A}$. Since p_λ is a $*$-preserving and submultiplicative seminorm on $\mathscr{A}$, it is easily proved that $\ker p_\lambda$ is a $*$-ideal of $\mathscr{A}$, and the quotient $*$-algebra $\mathscr{A}_\lambda := \mathscr{A}/_{\ker p_\lambda}$ is a normed $*$-algebra with the norm: $\|a + \ker p_\lambda\|_\lambda := p_\lambda(a)$, $a \in \mathscr{A}$. We denote by $\overline{\mathscr{A}_\lambda}$ the Banach $*$-algebra obtained by the completion of the normed $*$-algebra $\mathscr{A}_\lambda$. We now put

$$F(a + \ker p_\lambda) = f(a), \quad a \in \mathscr{A}.$$

By (5.2.8) F is a continuous positive linear functional on the normed $*$-algebra $\mathscr{A}_\lambda$, so that it can be extended to the continuous positive linear functional on the Banach $*$-algebra $\overline{\mathscr{A}_\lambda}$ and denote it by the same F. By Corollary 5.2.6 F is representable and

$$|F(X^*A^*AX)| \leqq \|A\|_\lambda^2 F(X^*X)$$

for all $A, X \in \overline{\mathscr{A}_\lambda}$. Hence we get

$$f(x^*a^*ax) \leqq p_\lambda(a)^2 f(x^*x)$$

for all $a, x \in \mathscr{A}$, which implies that f is representable and $\|\pi_f(a)\| \leqq p_\lambda(a)$ for all $a \in \mathscr{A}$.

(2) Let φ be a weakly continuous i.p.s. form on $\mathscr{A}$. For any $x \in \mathscr{A}$ the positive linear functional φ_x on $\mathfrak{A}$ is continuous, so that we can define in the same way as (1) the continuous positive linear functional $\bar{\varphi}_x$ on the Banach $*$-algebra $\overline{\mathscr{A}_\lambda}$ satisfying

$$\bar{\varphi}_x(a + \ker p_\lambda) = \varphi_x(a), \quad a \in \mathscr{A}.$$

By the Schwarz inequality for φ_x we can prove that

$$\begin{aligned}
|\varphi(hx, x)| &\leqq \varphi(x,x)^{\frac{1}{2}} \varphi(hx, hx)^{\frac{1}{2}} \\
&= \varphi(x,x)^{\frac{1}{2}} \varphi(h^2x, x)^{\frac{1}{2}} \\
&\leqq \varphi(x,x)^{\frac{1}{2}+\frac{1}{4}} \varphi(h^4x, x)^{\frac{1}{4}} \\
&\leqq \varphi(x,x)^{\sum_{k=1}^n (\frac{1}{2})^k} \varphi_x(h^{2^n})^{\frac{1}{2^n}} \\
&= \varphi(x,x)^{\sum_{k=1}^n (\frac{1}{2})^k} \tilde{\varphi}_x(h^{2^n} + \ker p_\lambda)^{\frac{1}{2^n}};
\end{aligned}$$

hence

$$\begin{aligned}
|\varphi(hx, x)| &\leqq \lim_{n\to\infty} \varphi(x,x)^{\sum_{k=1}^n (\frac{1}{2})^k} \tilde{\varphi}_x(h^{2^n} + \ker p_\lambda)^{\frac{1}{2^n}} \\
&= \varphi(x,x) \lim_{n\to\infty} \tilde{\varphi}_x(h^{2^n} + \ker p_\lambda)^{\frac{1}{2^n}} \\
&\leqq \varphi(x,x) \|h + \ker p_\lambda\|_\lambda \\
&= \varphi(x,x) p_\lambda(h)
\end{aligned}$$

for all $h \in \mathscr{A}_h$. Therefore $\|\pi_\varphi(h)\| \leqq p_\lambda(h)$ for all $h \in \mathscr{A}_h$, which implies that

$$\begin{aligned}
\|\pi_\varphi(a)\|^2 &= \|\pi_\varphi(a^*a)\| \\
&\leqq p_\lambda(a^*a) \\
&\leqq p_\lambda(a)^2
\end{aligned}$$

for all $a \in \mathscr{A}$. Thus φ is representable and $\|\pi_\varphi(a)\| \leqq p_\lambda(a)$ for all $a \in \mathscr{A}$. This completes the proof.

As for the equivalence of algebraic representability and representability of i.p.s. forms we have the following

Proposition 5.2.8 *Suppose that $\mathscr{A}$ is a barreled space. Then, every weakly continuous algebraically representable i.p.s. form on $\mathscr{A}$ is representable.*

Proof Let φ be any weakly continuous algebraically representable i.p.s. form on $\mathscr{A}$. For any $\varepsilon > 0$ put

$$U = \{a \in \mathscr{A};\quad \|\pi_\varphi(a)\| \leqq \varepsilon\}.$$

Then we have

$$U = \bigcap\{V_{x,y};\ x, y \in \mathscr{A}\quad \text{s.t.}\quad \|\lambda_\varphi(x)\| \leqq 1 \quad\text{and}\quad \|\lambda_\varphi(y)\| \leqq 1\},$$

where

$$\begin{aligned} V_{x,y} &:= \{a \in \mathscr{A};\ |(\pi_\varphi(a)\lambda_\varphi(x)|\lambda_\varphi(y))| \leqq \varepsilon\} \\ &= \{a \in \mathscr{A};\ |\varphi_{x,y}(a)| \leqq \varepsilon\}. \end{aligned}$$

By the weak continuity of φ, $\varphi_{x,y}$ is continuous, so that $V_{x,y} = \varphi_{x,y}^{-1}([-\varepsilon, \varepsilon])$ is a closed subset of $\mathscr{A}$. Therefore U is a closed, absolutely convex and absorbing subset of $\mathscr{A}$. Since $\mathscr{A}$ is a barreled locally convex space, it follows that U is a neighborhood of 0 in $\mathscr{A}$, which implies that π_φ is continuous. Thus φ is representable, which completes the proof.

Henceforth we treat only admissible i.p.s. forms and admissible positive linear functionals, and so we assume that i.p.s. forms and positive linear functionals are admissible without notice.

Notes The proof of Theorem 5.2.4 is an analogy with that of Theorem 3.12 in [5]. Proposition 5.2.8 is due to Theorem 4.6 in [5]. Locally m-convex $*$-algebras have been studied in [11, 12, 19], and studies of general locally convex $*$-algebras were done in [1, 2, 4, 5, 10, 27].

5.2.3 *Strongly Regular i.p.s. Forms*

In this subsection we investigate when an i.p.s. form on a $*$-algebra $\mathscr{A}$ is induced by a positive linear functional on $\mathscr{A}$. In Sect. 5.2.1 we have mentioned that this does not necessarily hold. Let $\mathscr{A}$ be a $*$-algebra. We first prepare the following notions:

Definition 5.2.9 Let φ be an i.p.s. form on $\mathscr{A}$. An i.p.s. form ψ on $\mathscr{A}$ is said to be φ-dominated if there exists a positive constant $\gamma > 0$ such that $\psi(a, a) \leqq \gamma\varphi(a, a)$ for all $a \in \mathscr{A}$, and then is denoted by $\psi \leqq \gamma\varphi$. A positive linear functional f on $\mathscr{A}$ is said to be φ-majorized if f^0 is φ-dominated and there exists a positive constant $\gamma > 0$ such that $|f(a)|^2 \leqq \gamma\varphi(a, a)$ for all $a \in \mathscr{A}$.

The following simple result is often useful.

Lemma 5.2.10 *Let φ and ψ be i.p.s. forms on $\mathscr{A}$ and f a positive linear functional on $\mathscr{A}$. Then*

(1) ψ is φ-dominated if and only if there exists a positive operator K_0 in $\pi_\varphi(\mathscr{A})'$ such that

$$\psi(a, b) = (K_0\lambda_\varphi(a)|\lambda_\varphi(b)) \quad for \quad all \quad a, b \in \mathscr{A};$$

(2) f is φ-majorized if and only if there exists an element K of $\mathfrak{A}_\varphi^\tau$ such that $\pi(K) \geqq 0$ and

$$f(a) = (\lambda_\varphi(a)|\lambda(K)) \quad for \quad all \quad a \in \mathscr{A}.$$

Proof

(1) Suppose that ψ is φ-dominated. Then, the map: $\lambda_\varphi(a) \rightarrow \lambda_\psi(a)$ can be extended to the continuous linear operator S_0 of $\mathcal{H}_\varphi$ into $\mathcal{H}_\psi$. Put $K_0 = S_0^* S_0$. Then it is easily shown that $K_0 \in \pi_\varphi(\mathscr{A})'$ and $\psi(a, b) = (K_0\lambda_\varphi(a)|\lambda_\varphi(b))$ for all $a, b \in \mathscr{A}$. The converse is trivial.

(2) Suppose that f is φ-majorized. Then, the functional: $\lambda_\varphi(a) \rightarrow f(a)$ can be extended to the continuous linear functional on $\mathcal{H}_\varphi$. By the Riesz theorem there exists an element g of $\mathcal{H}_\varphi$ such that $f(a) = (\lambda_\varphi(a)|g)$ for all $a \in \mathscr{A}$. By (1) there exists a positive operator K_0 in $\pi_\varphi(\mathscr{A})'$ such that $f(b^*a) = (\lambda_\varphi(a)|K_0\lambda_\varphi(b))$ for all $a, b \in \mathscr{A}$. Therefore $K_0\lambda_\varphi(b) = \pi_\varphi(b)g$ for all $b \in \mathscr{A}$, which implies that $K := (K_0, g, g^*) \in \mathfrak{A}_\varphi^\tau$ and $f(a) = (\lambda_\varphi(a)|\lambda(K))$ for all $a \in \mathscr{A}$. This completes the proof.

The following is the main consequence of this subsection.

Theorem 5.2.11 *Let φ be an i.p.s. form on $\mathscr{A}$. Then the following statements are equivalent:*

(i) $\varphi = f^0$ for some φ-majorized positive linear functional f on $\mathscr{A}$.
(ii) φ is extendable, that is, it can be extended to an i.p.s. form on the unitization $\mathscr{A}_e$ of $\mathscr{A}$.
(iii) $\varphi(a, b) = (\pi_\varphi(a)g|\pi_\varphi(b)g)$, $a, b \in \mathscr{A}$ for some $g \in \mathcal{H}_\varphi$.
(iv) $\lambda_\varphi(a) = \pi_\varphi(a)g$, $a \in \mathscr{A}$ for some $g \in \mathcal{H}_\varphi$.
(v) The map: $\pi_\varphi(a) \rightarrow \lambda_\varphi(a)$ is uniformly continuous.
(vi) $I \in \pi(\mathfrak{A}_\varphi^\tau)$.
(vii) $\pi_\varphi(\mathscr{A})' = \pi(\mathfrak{A}_\varphi^\tau)$.

Proof (i)⇒(iv) By Lemma 5.2.10 there exists an element g of $\mathcal{H}_\varphi$ such that

$$f(a) = (\lambda_\varphi(a)|g) \quad \text{for all} \quad a \in \mathscr{A} \tag{5.2.9}$$

Since $\varphi = f^0$, it follows from (5.2.9) that

$$\begin{aligned}(\lambda_\varphi(b)|\lambda_\varphi(a)) &= \varphi(b, a) \\ &= f(a^*b) \\ &= (\lambda_\varphi(b)|\pi_\varphi(a)g)\end{aligned}$$

for all $a, b \in \mathscr{A}$, which implies that $\lambda_\varphi(a) = \pi_\varphi(a)g$ for all $a \in \mathscr{A}$.
(iv)⇔(v) This follows from Lemma 2.2.1.
(iv)⇒(vii) Take an arbitrary $K_0 \in \pi_\varphi(\mathscr{A})'$, and put

$$K = (K_0, K_0 g, (K_0^* g)^*).$$

Then it is easily shown that $K \in \mathfrak{A}_\varphi^\tau$ and $K_0 = \pi(K) \in \pi(\mathfrak{A}_\varphi^\tau)$. Hence, $\pi_\varphi(\mathscr{A})' \subset \pi(\mathfrak{A}_\varphi^\tau)$. The converse inclusion is trivial.
(vii)⇒(vi) This is trivial.
(vi)⇒(iii) Since $I \in \pi(\mathfrak{A}_\varphi^\tau)$, there exist elements $g, h \in \mathcal{H}_\varphi$ such that $(I, g, h^*) \in \mathfrak{A}_\varphi^\tau$. Then $\lambda_\varphi(a) = \pi_\varphi(a)g$ for all $a \in \mathscr{A}$, and so

$$\varphi(a, b) = (\pi_\varphi(a)g|\pi_\varphi(b)g) \quad \text{for all} \quad a, b \in \mathscr{A}.$$

(iii)⇒(ii) Put

$$\varphi_e(a + \alpha e, b + \beta e) = \varphi(a, b) + \bar{\beta}(\pi_\varphi(a)g|g) + \alpha(g|\pi_\varphi(b)g) + \alpha\bar{\beta}\|g\|^2$$

for each $a, b \in \mathscr{A}$ and $\alpha, \beta \in \mathbb{C}$. Then it is easily shown that φ_e is an invariant sesquilinear form on $\mathscr{A}_e \times \mathscr{A}_e$ which is an extension of φ. Furthermore, by the inequality:

$$\varphi_e(a + \alpha e, a + \alpha e) = \|\pi_\varphi(a)g + \alpha g\|^2 \geqq 0,$$

φ_e is positive.
(ii)⇒(i) Let $\tilde{\varphi}$ be an i.p.s. form on $\mathscr{A}_e$ which is an extension of φ. Putting

$$f(a) = \tilde{\varphi}(a, e), \quad a \in \mathscr{A},$$

f is a positive hermitian linear functional on $\mathscr{A}$ satisfying

$$f^0(a, b) = f(b^*a) = \tilde{\varphi}(a, b) = \varphi(a, b)$$

for all $a,\ b \in \mathscr{A}$ and

$$|f(a)|^2 \leqq \tilde{\varphi}(a, a)\tilde{\varphi}(e, e) = f(a^*a)\tilde{\varphi}(e, e)$$

for all $a \in \mathscr{A}$. Hence φ is induced by the φ-majorized positive linear functional f. This completes the proof.

Definition 5.2.12 An i.p.s. form on $\mathscr{A}$ is called strongly regular if it satisfies one of the equivalent conditions in Theorem 5.2.11, and a positive linear functional f on $\mathscr{A}$ is called strongly regular if f^0 is strongly regular.

Example 5.2.13

(1) Let $\mathscr{A}$ be a $*$-algebra with indentity e. Then, every i.p.s. form φ on $\mathscr{A}$ is strongly regular. Indeed, the positive linear functional f on $\mathscr{A}$ defined by $f(a) = \varphi(a, e), a \in \mathscr{A}$ is φ-majorized and $f^0 = \varphi$.
(2) Let $\mathscr{A}$ be a locally convex $*$-algebra. A net $\{e_\alpha\}$ in $\mathscr{A}$ is said to be approximate identity of $\mathscr{A}$ if $\lim_\alpha e_\alpha a = \lim_\alpha a e_\alpha = a$ for all $a \in \mathscr{A}$, and it is called bounded if $\{e_\alpha\}$ and $\{e_\alpha^* e_\alpha\}$ are bounded subsets of $\mathscr{A}$. It is well known that every C^*-algebra, or more generally, locally convex GB^*-algebra has a bounded approximate identity [4]. For locally convex GB^*-algebras see [2, 10]. Every continuous positive linear functional f on a locally convex $*$-algebra $\mathscr{A}$ with a bounded approximate identity $\{e_\alpha\}$ is strongly representable. Indeed, this follows from

$$|f(a)|^2 = \lim_\alpha |f(e_\alpha a)|^2 \leqq \overline{\lim_\alpha} f(e_\alpha^* e_\alpha) f(a^*a)$$

for all $a \in \mathscr{A}$.

As for the strongly regularity of positive linear functionals we have the following

Theorem 5.2.14 *Let $\mathscr{A}$ be a $*$-algebra and f a positive linear functional on $\mathscr{A}$. Consider the following statements:*

(i) f is f^0-majorized, equivalently,

$$|f(a)|^2 \leqq \gamma f(a^*a), \quad a \in \mathscr{A}$$

for some positive constant γ.

(ii) $f(a) = (\pi_f(a)g|g)$, $a \in \mathscr{A}$ for some $g \in \mathcal{H}_f$.

(iii) f is extendable, that is, it can be extended to a positive linear functional on $\mathscr{A}_e$.

(iv) $\pi_f(a) \to f(a)$ is uniformly continuous.

(v) $\tau_f(a) \to f(a)$ is uniformly continuous.

(vi) f is strongly regular.

(vii) f^0 is extendable, that is, it can be extended to an i.p.s. form on $\mathscr{A}_e$.

(viii) $f(b^*a) = (\pi_f(b^*a)g|g)$, $a, b \in \mathscr{A}$ for some $g \in \mathcal{H}_f$.

(viv) $\lambda_f(a) = \pi_f(a)g$, $a \in \mathscr{A}$ *for some* $g \in \mathcal{H}_f$.

Then the following implications hold:

$$\begin{array}{ccccccccc} (i) & \Longleftrightarrow & (ii) & \Longleftrightarrow & (iii) & \Longleftrightarrow & (iv) & \Longleftrightarrow & (v) \\ & & & & \Downarrow & & & & \\ & & (vi) & \Longleftrightarrow & (vii) & \Longleftrightarrow & (viii) & \Longleftrightarrow & (viv). \end{array}$$

Proof (i)⇒(ii) By Lemma 5.2.10 there exists an element g of $\mathcal{H}_f$ such that $f(a) = (\lambda_f(a)|g)$ for all $a \in \mathscr{A}$, which implies that $\lambda_f(a) = \pi_f(a)g$ for all $a \in \mathscr{A}$.
(ii)⇒(iii) Put

$$f_e(a + \alpha e) = f(a) + \alpha \|g\|^2, \quad a \in \mathscr{A},\ \alpha \in \mathbb{C}.$$

Then since

$$\begin{aligned} f_e((a + \alpha e)^*(a + \alpha e)) &= f(a^*a + \alpha a^* + \bar{\alpha} a) + |\alpha|^2 \|g\|^2 \\ &= \|\pi_f(a)g + \alpha g\|^2 \geqq 0 \end{aligned}$$

for all $a \in \mathscr{A}$ and $\alpha \in \mathbb{C}$, it follows that f_e is a positive linear functional on $\mathscr{A}_e$ which is an extension of f.
(iii)⇒(i) Let f_e be a positive linear functional on $\mathscr{A}_e$ which is an extension of f. Then we have

$$\begin{aligned} |f(a)|^2 &= |f_e(a)|^2 \\ &\leqq f_e(e) f_e(a^*a) \\ &= f_e(e) f(a^*a) \end{aligned}$$

for all $a \in \mathscr{A}$. Hence, (i) holds. Thus, (i)–(iii) are equivalent. Furthermore, the equivalence of (i), (iv) and (v) follows from Lemma 2.2.2. By Theorem 5.2.11, (vi)–(viv) are equivalent, and (i)⇒(vi) is trivial. This completes the proof.

Remark 5.2.15

(1) Every strongly regular positive linear functional f on $\mathscr{A}$ is automatically hermitian.
(2) Suppose that f is strongly regular. Then, $f^0 = h^0$ for some f^0-majorized positive linear functional h on $\mathscr{A}$, but f is not necessarily f^0-majorized, that is, $f(b^*a) = h(b^*a)$ for all $a, b \in \mathscr{A}$, but $f \neq h$ in general.

Notes The mutual equivalence of (i), (iii), (iv), (vi) and (vii) in Theorem 5.2.11 is due to Theorem 4.4 in [16]. The equivalence of (i) and (ii) and the equivalence (vi) and (viv) are due to Corollary 4.5 in [16].

5.2.4 Decomposition of i.p.s. Forms

In this subsection we define the notions of regular i.p.s. forms and singular i.p.s. forms, and show that every i.p.s. form is decomposed into the regular i.p.s. form and the singular i.p.s. form using the projections defined in Sect. 4.1.

Let $\mathscr{A}$ be a $*$-algebra and φ an i.p.s. form on $\mathscr{A}$. Now we recall that $\tau_\varphi(\mathscr{A}) = \{(\pi_\varphi(a), \lambda_\varphi(a), \lambda_\varphi(a^*)^*);\ a \in \mathscr{A}\}$ and $\mathfrak{A}_\varphi = \overline{\tau_\varphi(\mathscr{A})}[\tau_u]$. We here use the following projections defined in Sect. 4.4.1:

$$P_{\mathfrak{A}_\varphi} := \text{Proj}\,[\pi(\mathfrak{A}_\varphi)\mathcal{H}_\varphi], \quad N_{\mathfrak{A}_\varphi} := \text{Proj}\,[\lambda(N(\mathfrak{A}_\varphi))],$$
$$F_{\mathfrak{A}_\varphi} := P_{\mathfrak{A}_\varphi}(I - N_{\mathfrak{A}_\varphi}), \quad S_{\mathfrak{A}_\varphi} := P_{\mathfrak{A}_\varphi} N_{\mathfrak{A}_\varphi},$$
$$Z_{\mathfrak{A}_\varphi} := (I - P_{\mathfrak{A}_\varphi})N_{\mathfrak{A}_\varphi}, \quad V_{\mathfrak{A}_\varphi} := (I - P_{\mathfrak{A}_\varphi})(I - N_{\mathfrak{A}_\varphi}).$$

The projection $P_{\mathfrak{A}_\varphi}$ is in $\pi_\varphi(\mathscr{A})' \cap \pi_\varphi(\mathscr{A})''$ and others are in $\pi_\varphi(\mathscr{A})'$, and

$$I = F_{\mathfrak{A}_\varphi} + N_{\mathfrak{A}_\varphi} + V_{\mathfrak{A}_\varphi} = F_{\mathfrak{A}_\varphi} + S_{\mathfrak{A}_\varphi} + Z_{\mathfrak{A}_\varphi} + V_{\mathfrak{A}_\varphi}.$$

Since

$$\begin{aligned}(V_{\mathfrak{A}_\varphi}\lambda_\varphi(a)|\lambda_\varphi(b)) &= \Big((I - P_{\mathfrak{A}_\varphi})P_{\mathfrak{A}_\varphi^\tau}\lambda_\varphi(a)|\lambda_\varphi(b)\Big) \\ &= \lim_\gamma \Big((I - P_{\mathfrak{A}_\varphi})\lambda_\varphi(a)|\pi(e'_\gamma)\lambda_\varphi(b)\Big) \\ &= \lim_\gamma \Big((I - P_{\mathfrak{A}_\varphi})\pi_\varphi(b^*)\lambda_\varphi(a)|\lambda(e'_\gamma)\Big) \\ &= 0\end{aligned}$$

for all $a, b \in \mathscr{A}$, we have $V_{\mathfrak{A}_\varphi} = 0$, where e'_γ is a net in $u_\pi(\mathfrak{A}_\varphi^\tau)$ such that $\pi(e'_\gamma)$ converges strongly to $P_{\mathfrak{A}_\varphi^\tau}$. Hence

$$I = F_{\mathfrak{A}_\varphi} + N_{\mathfrak{A}_\varphi} = F_{\mathfrak{A}_\varphi} + S_{\mathfrak{A}_\varphi} + Z_{\mathfrak{A}_\varphi}. \tag{5.2.10}$$

We now put, for $a, b \in \mathscr{A}$,

$$\begin{aligned}\varphi_r(a, b) &= (F_{\mathfrak{A}_\varphi}\lambda_\varphi(a)|\lambda_\varphi(b)), \\ \varphi_s(a, b) &= (N_{\mathfrak{A}_\varphi}\lambda_\varphi(a)|\lambda_\varphi(b)), \\ \varphi_{ns}(a, b) &= (S_{\mathfrak{A}_\varphi}\lambda_\varphi(a)|\lambda_\varphi(b)), \\ \varphi_n(a, b) &= (Z_{\mathfrak{A}_\varphi}\lambda_\varphi(a)|\lambda_\varphi(b))\end{aligned}$$

Using (5.2.1), we can prove the following

Theorem 5.2.16 *Let φ be an i.p.s. form on $\mathscr{A}$. Then φ_r, φ_s, φ_{ns} and φ_n are φ-dominated i.p.s. forms on $\mathscr{A}$, and φ is decomposed into*

$$\varphi = \varphi_r + \varphi_s = \varphi_r + \varphi_{ns} + \varphi_n.$$

Definition 5.2.17 An i.p.s. form φ on $\mathscr{A}$ is called regular (resp. singular, nondegenerate and singular, nilpotent) if $\varphi = \varphi_r$ (resp. $\varphi = \varphi_s$, $\varphi = \varphi_{ns}$, $\varphi = \varphi_n$).

The following example shows that there exist i.p.s. forms of each type.

Example 5.2.18 Let $\mathfrak{A}$ be a CT^*-algebra such that $\lambda(\mathfrak{A})$ is dense in $\mathcal{H}$. Put

$$\varphi(A, B) = (\lambda(A), \lambda(B)), \quad A, B \in \mathfrak{A}.$$

Then φ is an i.p.s. form on $\mathfrak{A}$, and $\mathfrak{A}_\varphi$ is isomorphic to $\mathfrak{A}$. If $\mathfrak{A}$ is regular (resp. singular, nondegenerate and singular, nilpotent), then φ is regular (resp. singular, nondegenerate and singular, nilpotent). For example, when $\mathfrak{A} = T^*(\mathcal{H})$, φ is nondegenerate and singular. When $\mathfrak{A} = N(\mathfrak{A})$, φ is nilpotent.

5.2.5 *Regularity and Singularity of i.p.s. Forms*

The following theorem as for the regularity of i.p.s. forms is the main consequence of this section.

Theorem 5.2.19 *Let $\mathscr{A}$ be a $*$-algebra and φ an i.p.s. form on $\mathscr{A}$. Consider the following statements:*

(i) *φ is regular.*
(ii) *There exists a net $\{f_\alpha\}$ of φ-majorized positive linear functionals on $\mathscr{A}$ such that $f_\alpha^0 \leqq \varphi$ for each α and $\lim_\alpha f_\alpha^0(a, b) = \varphi(a, b)$ for each $a, b \in \mathscr{A}$.*
(iii) *There exists a set $\{f_\alpha\}$ of φ-majorized positive linear functionals on $\mathscr{A}$ such that*

$$\varphi(a, b) = \sum_\alpha f_\alpha^0(a, b) \quad \text{for all} \quad a, b \in \mathscr{A}.$$

(iv) *There exists an element $\{g_\alpha\}$ of $\mathfrak{M}_\varphi$ such that*

$$\lambda_\varphi(a) = \sum_\alpha \pi_\varphi(a) g_\alpha \quad \text{for all} \quad a \in \mathscr{A},$$

where

$$\begin{aligned}\mathfrak{M}_\varphi &:= \mathfrak{M}_{\mathfrak{A}_\varphi}\\ &= \{\{g_\alpha\} \subset \mathcal{H};\quad (\pi_\varphi(a)g_\alpha|g_\beta) = 0,\quad \alpha \neq \beta\\ &\qquad and \quad \sum_\alpha \|\pi_\varphi(a)g_\alpha\|^2 \leqq \|\lambda_\varphi(a)\|^2 \ \ for \ \ all \ \ a \in \mathscr{A}\}.\end{aligned}$$

(v) $\mathfrak{A}_\varphi$ *is regular.*
(vi)

$$\|\lambda(A)\|^2 = \inf\left\{\sum_{n=1}^{\infty} \varphi(a_n, a_n);\quad \{a_n\} \in \mathfrak{F}(A)\right\},\quad A \in \mathfrak{A}_\varphi,$$

where

$$\begin{aligned}\mathfrak{F}(A) := \{\{a_n\} \in \mathscr{A};\ & \sum_{n=1}^{\infty} \|\pi_\varphi(a_n)\|^2 < \infty\\ & and \ \sum_{n=1}^{\infty} \pi_\varphi(a_n)^*\pi_\varphi(a_n) \geqq \pi(A)^*\pi(A)\}.\end{aligned}$$

Then

$$\begin{array}{c}(i) \Longleftrightarrow (ii) \Longleftrightarrow (iii) \Longleftrightarrow (iv) \Longleftrightarrow (v)\\ \Downarrow\\ (vi).\end{array}$$

Proof (i)⇔(v) This is trivial.
(v)⇒(iv) This follows from Theorem 4.2.6 and Theorem 4.4.4.
(iv)⇒(iii) Putting

$$f_\alpha(a) = (\pi_\varphi(a)g_\alpha|g_\alpha),\quad a \in \mathscr{A},$$

f_α is a φ-majorized positive linear functional on $\mathscr{A}$ such that $f_\alpha^0 \leqq \varphi$ and $\varphi(a, b) = \sum_\alpha f_\alpha^0(a, b)$ for all $a, b \in \mathscr{A}$.
(iii)⇒(ii) This is trivial.
(ii)⇒(v) By Lemma 5.2.10, there exists a net $\{K_\alpha\} \in \mathfrak{A}_\varphi^\tau$ satisfying $f_\alpha(a) = (\lambda_\varphi(a)|\lambda(K_\alpha))$ for any $a \in \mathscr{A}$ and $\pi(K_\alpha) \geqq 0$. Since $f_\alpha^0 \leqq \varphi$, it follows that

$$\begin{aligned}(\pi(K_\alpha)\lambda_\varphi(a)|\lambda_\varphi(a)) &= (\lambda_\varphi(a)|\pi(K_\alpha)\lambda_\varphi(a))\\ &= (\lambda_\varphi(a)|\pi_\varphi(a)\lambda(K_\alpha))\\ &= (\lambda_\varphi(a^*a)|\lambda(K_\alpha))\end{aligned}$$

$$
\begin{aligned}
&= f_\alpha(a^*a) \\
&= f_\alpha^0(a, a) \\
&\leqq \varphi(a, a) \\
&= (\lambda_\varphi(a)|\lambda_\varphi(a))
\end{aligned}
$$

for all $a \in \mathscr{A}$, which implies that

$$
0 \leqq \pi(K_\alpha) \leqq I \text{ and } f_\alpha(a) = (\lambda_\varphi(a)|\lambda(K_\alpha)) \tag{5.2.11}
$$

for all $a \in \mathscr{A}$. Take an arbitrary $A \in \mathfrak{A}_\varphi$. Then we can take a sequence $\{a_n\}$ in $\mathscr{A}$ such that $\lim_{n\to\infty} \|\tau_\varphi(a_n) - A\| = 0$, namely

$$
\lim_{n\to\infty} \|\pi_\varphi(a_n) - \pi(A)\| = 0, \quad \lim_{n\to\infty} \|\lambda_\varphi(a_n) - \lambda(A)\| = 0, \quad \lim_{n\to\infty} \|\lambda_\varphi(a_n^*) - \lambda(A^\sharp)\| = 0. \tag{5.2.12}
$$

Hence it follows from assumption (ii) and (5.2.11) that

$$
\begin{aligned}
\|\lambda_\varphi(a)\|^2 = \varphi(a, a) &= \lim_\alpha(\lambda_\varphi(a^*a)|\lambda(K_\alpha)) \\
&= \lim_\alpha(\pi(K_\alpha)\lambda_\varphi(a)|\lambda_\varphi(a)) \\
&\leqq \lim_\alpha \|\pi(K_\alpha)\lambda_\varphi(a)\| \|\lambda_\varphi(a)\| \\
&\leqq \|\lambda_\varphi(a)\|^2
\end{aligned}
$$

for all $a \in \mathscr{A}$. Therefore

$$
\lim_\alpha \|\pi(K_\alpha)\lambda_\varphi(a)\| = \|\lambda_\varphi(a)\| \tag{5.2.13}
$$

for all $a \in \mathscr{A}$. Since

$$
\begin{aligned}
\big|\|\pi(A)\lambda(K_\alpha)\| - \|\lambda(A)\|\big| &\leqq \|\pi(K_\alpha)\lambda(A) - \pi(K_\alpha)\lambda_\varphi(a_n)\| + \big|\|\pi(K_\alpha)\lambda_\varphi(a_n)\| - \|\lambda_\varphi(a_n)\|\big| \\
&\quad + \big|\|\lambda_\varphi(a_n)\| - \|\lambda(A)\|\big| \\
&\leqq \|\lambda(A) - \lambda_\varphi(a_n)\| + \big|\|\pi(K_\alpha)\lambda_\varphi(a_n)\| - \|\lambda_\varphi(a_n)\|\big| \\
&\quad + \big|\|\lambda_\varphi(a_n)\| - \|\lambda(A)\|\big|
\end{aligned}
$$

for all α and $n \in \mathbb{N}$, it follows from (5.2.13) that

$$
\lim_\alpha \big|\|\pi(A)\lambda(K_\alpha)\| - \|\lambda(A)\|\big| \leqq 2\|\lambda(A) - \lambda_\varphi(a_n)\|
$$

for all $n \in \mathbb{N}$, and by (5.2.12) that

$$\lim_{\alpha} \|\pi(A)\lambda(K_\alpha)\| = \|\lambda(A)\|.$$

Similarly, we get

$$\lim_{\alpha} \|\pi(A)^*\lambda(K_\alpha)\| = \|\lambda(A^\sharp)\|,$$

which prove that the map: $A \in \mathfrak{A}_\varphi \to \pi(A) \in B(\mathcal{H}_\varphi)$ is injective. By Theorem 4.2.6 $\mathfrak{A}_\varphi$ is semisimple, and it is regular since $\lambda(\mathfrak{A}_\varphi)$ is dense in $\mathcal{H}_\varphi$. Thus, (i)–(v) are equivalent.
(v)⇒(vi) This follows from Theorem 4.4.2.
This completes the proof.

As for the singularity of i.p.s. forms we have the following

Theorem 5.2.20 *Let φ be an i.p.s. form on $\mathscr{A}$. Then the following statements (1), (2) and (3) hold:*

(1) The following are equivalent:

(i) φ is singular.
(ii) There does not exist any φ-majorized positive linear functional f on $\mathscr{A}$ such that $f^0 \neq 0$.
(iii) $\mathfrak{A}_\varphi$ is singular.
(iv) Put

$$s_\varphi(\mathscr{A}) = \{\{a_n\} \subset \mathscr{A};\ \ \pi_\varphi(a_n) \to 0, \text{ uniformly and } \{\lambda_\varphi(a_n)\} \text{ and } \{\lambda_\varphi(a_n^*)\} \text{ are Cauchy sequences in } \mathcal{H}_\varphi\}.$$

Then $\lambda_\varphi(s_\varphi(\mathscr{A})) := \{\lim_{n\to\infty} \lambda_\varphi(a_n);\ \{a_n\} \in s_\varphi(\mathscr{A})\}$ is dense in $\mathcal{H}_\varphi$.

(2) The following are equivalent:

(i) φ is nondegenerate and singular.
(ii) $\lambda_\varphi(\mathscr{A}^2)$ is dense in $\mathcal{H}_\varphi$ and there does not exist any φ-majorized positive linear functional f on $\mathscr{A}$ such that $f^0 \neq 0$.
(iii) $\mathfrak{A}_\varphi$ is nondegenerate and singular.
(iv) For any $x, y \in \mathcal{H}_\varphi$ there exists a sequence $\{a_n\}$ in $\mathscr{A}^2$ such that $\tau_\varphi(a_n) = (\pi_\varphi(a_n), \lambda_\varphi(a_n), \lambda_\varphi(a_n^*)^*) \underset{n\to\infty}{\to} (0, x, y^*)$, uniformly.

(3) The following are equivalent:

(i) φ is nilpotent.
(ii) $\varphi(ab, c) = 0$ for all $a, b, c \in \mathscr{A}$.
(iii) $\tau_\varphi(\mathscr{A}) = N(\tau_\varphi(\mathscr{A}))$.

Proof

(1) The equivalence of (i), (ii) and (iii) follows from Theorem 4.1.7 and Lemma 5.2.10. The equivalence of (iii) and (iv) follows from Theorem 4.4.6, (1).
(2) Since $\lambda_\varphi(\mathscr{A})$ is dense in $\mathcal{H}_\varphi$, it is clear that φ is nondegenerate if and only if $\lambda_\varphi(\mathscr{A}^2)$ is dense in $\mathcal{H}_\varphi$. Hence it follows from (1) that (i), (ii) and (iii) are equivalent. The equivalence (iii) and (iv) follows from Theorem 4.4.6, (2)
(3) This follows from Theorem 4.4.6, (3).

This completes the proof.

5.3 Positive Definite Generalized Functions in Lie Groups

In this section we apply the theory of the observable algebras to positive definite generalized functions in a separable Lie group. Let G be a separable Lie group, dx the left invariant measure on G and Δ the modular function on G. Then Δ is continuous on G and

$$\int f(ax)dx = \int f(x)dx, \tag{5.3.1}$$

$$\int f(xa)dx = \Delta(a^{-1}) \int f(x)dx, \tag{5.3.2}$$

$$\int f(x^{-1})dx = \int f(x)\,\Delta\,(x^{-1})dx \tag{5.3.3}$$

for every integrable function f on G. Let $\mathcal{D}(G)$ denote the space of all infinitely differentiable functions on G with compact supports. The locally convex topology on $\mathcal{D}(G)$ is the strongest locally convex topology under which all the linear mappings $l_K : f \in \mathcal{D}_K(G) \to f \in \mathcal{D}(G)$ are continuous, where K is any compact subset of G and $\mathcal{D}_K(G) := \{f \in \mathcal{D}(G);\ \operatorname{supp} f \subset K\}$ is a Fréchet space under the topology defined by a family of seminorms:

$$p_{K,m}(f) := \sup\{|f^{(k)}(x)|;\ \ k \leqq m,\ x \in K\}, \quad m \in \mathbb{N}.$$

A base of neighborhoods for this topology is formed by the set of all absolutely convex subsets U of $\mathcal{D}(G)$ such that for any K, $l_K^{-1}(U)$ is a neighborhood in $\mathcal{D}_K(G)$. The locally convex space $\mathcal{D}(G) = \cup_K \mathcal{D}_K(G)$ is called the inductive limit of the Fréchet spaces $\mathcal{D}_K(G)$ by the mapping l_K. Furthermore, $\mathcal{D}(G)$ is a locally convex $*$-algebra equipped with the multiplication $f \circ g$ and the involution $f \to f^*$:

$$(f \circ g)(a) := \int f(x)g(x^{-1}a)dx,$$

$$f^*(a) := \Delta(a^{-1})\overline{f(a^{-1})},$$

which is called the *Schwartz group algebra*. A continuous linear functional on $\mathcal{D}(G)$ is called a *generalized function in* G. A locally integrable function p on G is identical with the generalized function $\hat{p}$ in G defined by

$$\hat{p}(f) = \int f(a)p(a)da, \quad f \in \mathcal{D}(G).$$

A generalized function μ in G is called *positive definite* if

$$\mu(f^* \circ f) \geqq 0 \quad \text{for all} \quad f \in \mathcal{D}(G).$$

Now we can prove the following fundamental result.

Theorem 5.3.1 *Every positive definite generalized function μ in G is a representable positive linear functional on the Schwartz group algebra $\mathcal{D}(G)$. Let $(\pi_\mu, \lambda_\mu, \mathcal{H}_\mu)$ be the GNS-construction for μ. Then*

$$\|\pi_\mu(f)\| \leqq \|f\|_1 := \int |f(a)|da$$

for all $f \in \mathcal{D}(G)$.

Proof For $a \in G$ and $f \in \mathcal{D}(G)$ we define an element f_a of $\mathcal{D}(G)$ by

$$f_a(x) = f(a^{-1}x), \quad x \in G.$$

Using (5.3.1)–(5.3.3), we can show that for any $f,\ g \in \mathcal{D}(G)$ and $a \in G$,

$$\begin{aligned}
(f^* \circ g_a)(x) &= \int f^*(y)g_a(y^{-1}x)dy \\
&= \int \Delta(y^{-1})\overline{f(y^{-1})}g(a^{-1}y^{-1}x)dy \\
&= \int \overline{f(y)}g(a^{-1}yx)dy \\
&= \int \overline{f(ay)}g(yx)dy
\end{aligned}$$

$$
\begin{aligned}
&= \int \overline{f_{a^{-1}}(y)} g(yx)dy \\
&= \int \overline{f_{a^{-1}}(y^{-1})} g(y^{-1}x) \triangle (y^{-1})dy \\
&= \int (f_{a^{-1}})^*(y) g(y^{-1}x)dy \\
&= \left((f_{a^{-1}})^* \circ g\right)(x)
\end{aligned}
$$

for all $x \in G$. Thus, we have

$$
f^* \circ g_a = (f_{a^{-1}})^* \circ g. \tag{5.3.4}
$$

Let μ be a positive definite generalized function in G. Then it follows from (5.3.4) and the Schwarz inequality for μ that

$$
\begin{aligned}
|\mu(g_b^* \circ g_a)| &= |\mu(g_{a^{-1}b}^* \circ g)| \\
&\leqq \mu(g_{a^{-1}b}^* \circ g_{a^{-1}b})^{\frac{1}{2}} \mu(g^* \circ g)^{\frac{1}{2}} \\
&= \mu(g^* \circ g)
\end{aligned} \tag{5.3.5}
$$

for all $a,\ b \in G$ and $g \in \mathcal{D}(G)$. We define a seminorm r on $\mathcal{D}(G)$ by

$$
r(f) = \mu(f^* \circ f)^{\frac{1}{2}}, \quad f \in \mathcal{D}(G).
$$

Then, since r is bounded on each bounded subset of $\mathcal{D}(G)$ and $\mathcal{D}(G)$ is a bornological space, it follows that r is continuous, which implies by (5.3.4) and (5.3.5) that

$$
\begin{aligned}
\mu^0(f \circ g, f \circ g) &= \mu^0\left(\int f(a) g_a(x)da, \int f(b) g_b(x)db\right) \\
&= \int\int f(a)\overline{f(b)}\mu^0(g_a, g_b)dadb \\
&= \int\int f(a)\overline{f(b)}\mu(g_b^* \circ g_a)dadb \\
&= \int\int f(a)\overline{f(b)}\mu(g_{a^{-1}b}^* \circ g)dadb \\
&\leqq \int\int |f(a)\overline{f(b)}|\mu(g^* \circ g)dadb \\
&= \|f\|_1^2 \mu(g^* \circ g)
\end{aligned}
$$

for all $f,\ g \in \mathcal{D}(G)$. Furthermore, since the L^1-norm $\|\cdot\|_1$ is a continuous norm on $\mathcal{D}(G)$, it follows that μ is representable, and the GNS representation by μ satisfies that $\|\pi_\mu(f)\| \leqq \|f\|_1$ for all $f \in \mathcal{D}(G)$. This completes the proof.

To investigate the strong regularity of positive definite generalized functions in G we prepare the following

Lemma 5.3.2 *Let μ be a positive definite generalized function in G and $(\pi_\mu, \lambda_\mu, \mathcal{H}_\mu)$ the GNS-represetation for μ. Then there exists a strongly continuous unitary representation $\{U_a;\ a \in G\}$ of G on $\mathcal{H}_\mu$ such that*

$$\pi_\mu(f) = \int f(a)U_a da, \quad f \in \mathcal{D}(G).$$

Proof Let $\{V_n;\ n \in \mathbb{N}\}$ be a base of open neighborhoods of the unit e of G, and for any n take an element $\delta_n \geqq 0$ of $\mathcal{D}(G)$ which vanishes outside of V_n and $\delta_n(e) = 1$. Then $\{\delta_n\}$ is an approximate identity of $\mathcal{D}(G)$, which called a δ-sequence in $\mathcal{D}(G)$. Furthermore, it follows from Theorem 5.3.1 that π_μ is a continuous nondegenerate $*$-representation of $\mathcal{D}(G)$ on the Hilbert space $\mathcal{H}_\mu$. Hence

$$\text{Span of } \{\pi_\mu(f)\xi;\ f \in \mathcal{D}(G),\ \xi \in \mathcal{H}_\mu\} \text{ is dense in } \mathcal{H}_\mu. \tag{4.3.5a}$$

Let $a \in G$. By (5.3.4) we have

$$g_a^* \circ f_a = ((g_a)_{a^{-1}})^* \circ f = g^* \circ f$$

for all $f,\ g \in \mathcal{D}(G)$, so that

$$\begin{aligned}(\pi_\mu(f)\xi|\pi_\mu(g)\eta) &= (\pi_\mu(g^* \circ f)\xi|\eta)\\ &= (\pi_\mu(g_a^* \circ f_a)\xi|\eta)\\ &= (\pi_\mu(f_a)\xi|\pi_\mu(g_a)\eta)\end{aligned}$$

for all $\xi,\ \eta \in \mathcal{H}_\mu$. Therefore, by (5.3.5), we can define a one-parameter unitary set $\{U_a;\ a \in G\}$ on $\mathcal{H}_\mu$ by

$$U_a\pi_\mu(f)\xi = \pi_\mu(f_a)\xi$$

for $f \in \mathcal{D}(G)$ and $\xi \in \mathcal{H}_\mu$. Since π_μ is strongly continuous and $a \in G \to f_a \in \mathcal{D}(G)$ is continuous, it follows that $a \in G \to U_a \in B(\mathcal{H}_\mu)$ is strongly continuous. Since $(f_b)_a = f_{ab}$ for all $a,\ b \in G$ and $f \in \mathcal{D}(G)$, we get that

$$\begin{aligned}U_aU_b\pi_\mu(f)\xi = U_a\pi_\mu(f_b)\xi &= \pi_\mu((f_b)_a)\xi\\ &= \pi_\mu(f_{ab})\xi\\ &= U_{ab}\pi_\mu(f)\xi\end{aligned}$$

for all $\xi \in \mathcal{H}_\mu$, which implies by (5.3.5) that $U_a U_b = U_{ab}$ and $U_{a^{-1}} = U_a^{-1}$. Thus $\{U_a;\ a \in G\}$ is a strongly continuous unitary representation of G on $\mathcal{H}_\mu$. Since

$$\begin{aligned}\int f(a)U_a\pi_\mu(g)\xi da &= \int f(a)\pi_\mu(g_a)\xi da\\ &= \pi_\mu\left(\int f(a)g_a da\right)\xi\\ &= \pi_\mu(f \circ g)\xi\\ &= \pi_\mu(f)\pi_\mu(g)\xi\end{aligned}$$

for all $f,\ g \in \mathcal{D}(G)$ and $\xi \in \mathcal{H}_\mu$, it follows from (5.3.5) that

$$\pi_\mu(f) = \int f(a)U_a da, \quad f \in \mathcal{D}(G).$$

This completes the proof.

We next characterize the strong regularity of positive definite generalized functions on G.

Theorem 5.3.3 *Let μ be a positive definite generalized function in G. Then the following statements are equivalent:*

(i) μ is strongly regular.
(ii) μ is induced by a continuous positive definite function p on G, that is,

$$\mu(f) = \int f(a)p(a)da, \quad f \in \mathcal{D}(G).$$

Proof (i)⇒(ii) Since μ is strongly regular, by Theorem 5.2.14 there exists an element ξ of $\mathcal{H}_\mu$ such that

$$\lambda_\mu(f) = \pi_\mu(f)\xi \quad \text{for all} \quad f \in \mathcal{D}(G).$$

Hence it follows from Lemma 5.3.2 that

$$\begin{aligned}\mu(g^* \circ f) &= (\lambda_\mu(f)|\lambda_\mu(g))\\ &= (\pi_\mu(f)\xi|\pi_\mu(g)\xi)\\ &= (\pi_\mu(g^* \circ f)\xi|\xi)\\ &= \int (g^* \circ f)(a)(U_a\xi|\xi)da\end{aligned}$$

for all $f,\ g \in \mathcal{D}(G)$. By taking δ-sequence, we get

$$\mu(f) = \int f(a)(U_a\xi|\xi)da$$

for all $f \in \mathcal{D}(G)$, and $p(a) := (U_a\xi|\xi)$, $a \in G$ is a continuous positive definite function on G. Thus μ is induced by the continuous positive definite function p on G.

(ii)$\Rightarrow$(i) It is well known that for every continuous positive definite function p on G there exists a strongly continuous unitary representation U_a of G on a Hilbert space $\mathcal{H}$ such that

$$p(a) = (U_a\xi|\xi), \quad a \in G$$

for some $\xi \in \mathcal{H}$. Then we can define a strongly continuous $*$-representation π of $\mathcal{D}(G)$ on $\mathcal{H}$ by

$$\pi(f) = \int f(a)U_a da, \quad f \in \mathcal{D}(G),$$

and then

$$\begin{aligned} |\mu(f)|^2 &= \left|\left(\int f(a)(U_a\xi|\xi)da\right)\right|^2 \\ &= |(\pi(f)\xi|\xi)|^2 \\ &\leqq \|\xi\|^2\|\pi(f)\xi\|^2 \\ &= \|\xi\|^2\mu(f^* \circ f) \end{aligned}$$

for all $f \in \mathcal{D}(G)$. Hence μ is strongly regular. This completes the proof.

Let μ be a positive definite generalized function in G, μ^0 the i.p.s. form on $\mathcal{D}(G)$ induced by μ and $\mathfrak{A}_\mu$ the CT^*-algebra which is the uniform closure of the T^*-algebra

$$\tau_\mu(\mathcal{D}(G)) := \{(\pi_\mu(f), \lambda_\mu(f), \lambda_\mu(f^*)^*); \quad f \in \mathcal{D}(G)\}.$$

By Theorem 5.2.16, μ^0 is decomposed into the regular part μ_r^0 and the singular part μ_s^0 of μ^0:

$$\mu^0 = \mu_r^0 + \mu_s^0,$$

where

$$\mu_r^0(f, g) := (F_{\mathfrak{A}_\mu}\lambda_\mu(f)|\lambda_\mu(g)),$$
$$\mu_s^0(f, g) := (N_{\mathfrak{A}_\mu}\lambda_\mu(f)|\lambda_\mu(g)), \quad f, g \in \mathcal{D}(G).$$

If $\mu^0 = \mu_r^0$ (resp. $\mu^0 = \mu_s^0$), then μ is called *regular* (resp. *singular*).
For the regularity of positive definite generalized functions in G we have the following

Theorem 5.3.4 *Let μ be a positive definite generalized function in G. Then the following statements are equivalent:*

(i) μ is regular.
(ii) The map: $\pi_\mu(f) \in \pi_\mu(\mathcal{D}(G)) \to \lambda_\mu(f) \in \mathcal{H}_\mu$ is uniformly closable.
(iii) There exists a set Λ of positive definite functions p on G such that

$$\mu(f^* \circ f) = \sum_{p \in \Lambda} \hat{p}(f^* \circ f)$$

for all $f \in \mathcal{D}(G)$.

Proof This follows from Theorems 5.2.19 and 5.3.2.

For the singularity of positive definite generalized functions in G we have the following

Theorem 5.3.5 *Let μ be a positive definite generalized function in G. Then the following statements are equivalent:*

(i) μ is singular.
(ii) For any $f \in \mathcal{D}(G)$ there exists a sequence $\{f_n\}$ in $\mathcal{D}(G)$ such that $\lim_{n\to\infty} \|\pi_\mu(f_n)\| = 0$ and $\lim_{n\to\infty} \|\lambda_\mu(f_n) - \lambda_\mu(f)\| = 0$.
(iii) There does not exist any positive definite function p on G such that $(\hat{p})^0 \neq 0$.

Proof This follows from Theorems 5.2.20 and 5.3.2.

Notes Almost all the results of this section are due to Section 5 in the Tomita unpublished paper [42]. In particular, Theorem 5.2.20 is due to Theorem 5.5 and Theorem 5.3.2 in [42]. Tomita studied non-commutative Fourier analysis which has closely related to the study of observable algebras. We here note it roughly. He considered to generalize the Bochner-Schwartz theorem (Every positive definite generalized function on the additive group $\mathbb{R}^n$ is the adjoint Fourier transform of a tempered Radon measure in $E(\mathbb{R}^n)$, Theorem 3 in 3.3 in [13]), and obtained that a positive definite generalized function in G becomes the adjoint Fourier transform of a certain tempered abelian measure if and only if it is regular. For detail, refer the reader to Section 5 in [42] and [13, 14].

5.4 Weights in C^*-Algebras

In this section we study weights in C^*-algebras using the theory of CT^*-algebras. Let $\mathscr{A}$ be a C^*-algebra and $\mathscr{A}_+ := \{a \in \mathscr{A};\ a \geqq 0\}$ the positive cone of $\mathscr{A}$. A functional φ of $\mathscr{A}_+$ into $[0, \infty]$ is called a *weight* in $\mathscr{A}$ if

(i) $\varphi(x + y) = \varphi(x) + \varphi(y), x, y \in \mathscr{A}_+$;
(ii) $\varphi(\alpha x) = \alpha\varphi(x), \alpha \geqq 0, x \in \mathscr{A}_+$,

where $o \cdot \infty = 0$.
Let φ be a weight in $\mathscr{A}$. We put

$$\begin{aligned} D(\varphi) &= \{x \in \mathscr{A}_+;\quad \varphi(x) < \infty\}, \\ \mathfrak{N}_\varphi &= \{x \in \mathscr{A};\quad \varphi(x^*x) < \infty\}, \\ \mathfrak{M}_\varphi &= \mathfrak{N}_\varphi^* \mathfrak{N}_\varphi, \\ \mathscr{A}_\varphi &= \mathfrak{N}_\varphi \cap \mathfrak{N}_\varphi^*. \end{aligned}$$

Then it is easily shown that

- $\mathfrak{N}_\varphi$ is a left ideal of $\mathscr{A}$,
- $D(\varphi) = \{x \in \mathscr{A}_+;\quad x^{\frac{1}{2}} \in \mathscr{A}_\varphi\}$,
- $\mathfrak{M}_\varphi = \mathscr{A}_\varphi^* \mathscr{A}_\varphi =$ linear span of $D(\varphi)$,

and φ can be extended uniquely to the positive linear functional on $\mathfrak{M}_\varphi$, which is denoted by the same φ. We can construct the GNS-construction $(\pi_\varphi, \lambda_\varphi, \mathcal{H}_\varphi)$ for φ as follows: Put

$$\begin{aligned} N_\varphi &= \{x \in \mathfrak{N}_\varphi;\quad \varphi(x^*x) = 0\}, \\ \lambda_\varphi(x) &= x + N_\varphi \in \mathfrak{N}_\varphi/_{N_\varphi}, \quad x \in \mathfrak{N}_\varphi. \end{aligned}$$

Then, $\lambda_\varphi(\mathfrak{N}_\varphi) := \{\lambda_\varphi(x);\ x \in \mathfrak{N}_\varphi\}$ is a pre-Hilbert space with inner product:

$$(\lambda_\varphi(x)|\lambda_\varphi(y)) = \varphi(y^*x), \quad x, y \in \mathfrak{N}_\varphi.$$

We denote by $\mathcal{H}_\varphi$ the Hilbert space obtained by the completion of $\lambda_\varphi(\mathfrak{N}_\varphi)$. Since $\mathfrak{N}_\varphi$ is a left ideal of $\mathscr{A}$, we can define a continuous $*$-representation π_φ of $\mathscr{A}$ on $\mathcal{H}_\varphi$ by

$$\pi_\varphi(a)\lambda_\varphi(x) = \lambda_\varphi(ax), \quad a \in \mathscr{A},\ x \in \mathfrak{N}_\varphi.$$

We next define the CT^*-algebra $\mathfrak{A}_\varphi$ constructed from φ. For any $x \in \mathscr{A}_\varphi$ we write

$$\tau_\varphi(x) = (\pi_\varphi(x), \lambda_\varphi(x), \lambda_\varphi(x^*)^*).$$

Then $\tau_\varphi(\mathscr{A}_\varphi)$ is a T^*-algebra on $\mathcal{H}_\varphi$, and denote by $\mathfrak{A}_\varphi$ the CT^*-algebra on $\mathcal{H}_\varphi$ obtained by the uniform closure of the T^*-algebra $\tau_\varphi(\mathscr{A}_\varphi)$. We here remark that $\lambda(\mathfrak{A}_\varphi)$ is not necessarily dense in $\mathcal{H}_\varphi$. Now, put

$$\varphi_r(x) = \begin{cases} (F_{\mathfrak{A}_\varphi}\lambda_\varphi(x^{\frac{1}{2}})|\lambda_\varphi(x^{\frac{1}{2}})) , & x \in D(\varphi) \\ \infty , & x \notin D(\varphi) \end{cases}$$

$$\varphi_s(x) = \begin{cases} (N_{\mathfrak{A}_\varphi}\lambda_\varphi(x^{\frac{1}{2}})|\lambda_\varphi(x^{\frac{1}{2}})) , & x \in D(\varphi) \\ \infty , & x \notin D(\varphi). \end{cases}$$

Then we have the following

Proposition 5.4.1 *Let φ be a weight in a C^*-algebra $\mathscr{A}$. Then φ_r and φ_s are weights in $\mathscr{A}$ with $D(\varphi_r) = D(\varphi_s) = D(\varphi)$ and $\varphi = \varphi_r + \varphi_s$.*

Proof Take arbitrary $x, y \in D(\varphi)$. Since $D(\varphi) = \{x \in \mathscr{A}_+; x^{\frac{1}{2}} \in \mathscr{A}_\varphi\}$, we have

$$\begin{aligned}
\varphi_r(x+y) &= \lim_\gamma \left(P_{\mathfrak{A}_\varphi}\pi(e'_\gamma)\lambda_\varphi((x+y)^{\frac{1}{2}})|\pi(e'_\gamma)\lambda_\varphi((x+y)^{\frac{1}{2}})\right) \\
&= \lim_\gamma \left(P_{\mathfrak{A}_\varphi}\pi_\varphi((x+y)^{\frac{1}{2}})\lambda(e'_\gamma)|\pi_\varphi((x+y)^{\frac{1}{2}})\lambda(e'_\gamma)\right) \\
&= \lim_\gamma \left(\pi_\varphi(x+y)\lambda(e'_\gamma)|\lambda(e'_\gamma)\right) \\
&= \lim_\gamma \left(\pi_\varphi(x)\lambda(e'_\gamma)|\lambda(e'_\gamma)\right) + \lim_\gamma \left(\pi_\varphi(y)\lambda(e'_\gamma)|\lambda(e'_\gamma)\right) \\
&= \lim_\gamma \left(P_{\mathfrak{A}_\varphi}\pi(e'_\gamma)\lambda_\varphi(x^{\frac{1}{2}})|\pi(e'_\gamma)\lambda_\varphi(x^{\frac{1}{2}})\right) \\
&\quad + \lim_\gamma \left(P_{\mathfrak{A}_\varphi}\pi(e'_\gamma)\lambda_\varphi(y^{\frac{1}{2}})|\pi(e'_\gamma)\lambda_\varphi(y^{\frac{1}{2}})\right) \\
&= \left(F_{\mathfrak{A}_\varphi}\lambda_\varphi(x^{\frac{1}{2}})|\lambda_\varphi(x^{\frac{1}{2}})\right) + \left(F_{\mathfrak{A}_\varphi}\lambda_\varphi(y^{\frac{1}{2}})|\lambda_\varphi(y^{\frac{1}{2}})\right) \\
&= \varphi_r(x) + \varphi_r(y),
\end{aligned}$$

where $\{e'_\gamma\}$ is a net in $\mathfrak{A}^\tau_\varphi$ such that $\pi(e'_\gamma) \to P_{\mathfrak{A}^\tau_\varphi}$, strongly. It is trivial that $\varphi_r(\alpha x) = \alpha\varphi_r(x)$ for all $\alpha \geqq 0$ and $x \in D(\varphi)$. Suppose that $x \notin D(\varphi)$ or $y \notin D(\varphi)$. Then, $\varphi_r(x) = \infty$ or $\varphi_r(y) = \infty$. Furthermore, since $\varphi(x+y) = \infty$, we have $x + y \notin D(\varphi)$, and so $\varphi_r(x+y) = \infty$. Hence, $\varphi_r(x+y) = \varphi_r(x) + \varphi_r(y) = \infty$. It is immediately proved that for any $x \notin D(\varphi)$

$$\varphi_r(\alpha x) = \alpha\varphi_r(x) = \begin{cases} 0 , & \alpha = 0 \\ \infty , & \alpha > 0. \end{cases}$$

Thus, φ_r is a weight in $\mathscr{A}$. Take an arbitrary $x \in D(\varphi)$. The equality:

$$\begin{aligned}\varphi_s(x) &= \left((I - P_{\mathfrak{A}_\varphi^\tau})\lambda_\varphi(x^{\frac{1}{2}})|\lambda_\varphi(x^{\frac{1}{2}})\right) \\ &= \varphi(x) - \varphi_r(x)\end{aligned}$$

implies

$$\begin{aligned}\varphi_s(x+y) &= \varphi_s(x) + \varphi_s(y), \\ \varphi_s(\alpha x) &= \alpha\varphi_s(x)\end{aligned}$$

for all $x, y \in D(\varphi)$ and $\alpha \geqq 0$. Suppose that $x \notin D(\varphi)$ or $y \notin D(\varphi)$. Then, $\varphi_s(x) = \infty$ or $\varphi_s(y) = \infty$. Furthermore, $\varphi_s(x+y) = \infty$ because $x + y \notin D(\varphi)$. Hence, $\varphi_s(x+y) = \varphi_s(x) + \varphi_s(y) = \infty$. It is trivial that for any $x \notin D(\varphi)$

$$\varphi_s(\alpha x) = \alpha\varphi_s(x) = \begin{cases} 0\ , & \alpha = 0 \\ \infty\ , & \alpha > 0. \end{cases}$$

Thus, φ_s is a weight in $\mathscr{A}$, and $\varphi = \varphi_r + \varphi_s$. This completes the proof.

The weight φ_r (resp. φ_s) in $\mathscr{A}$ is called the *regular* (resp. *singular*) *part* of φ. If $\varphi = \varphi_r$, then φ is called *regular*, and if $\varphi = \varphi_s$, then φ is called *singular*. We now give some examples of weights in $\mathscr{A}$.

Example 5.4.2 Let $\mathscr{A}$ be a C^*-algebra.

(1) Let $\{f_\alpha\}_{\alpha\in\mathbb{A}}$ be a net of positive linear functionals on $\mathscr{A}$. Putting

$$\left(\sup_{\alpha\in\mathbb{A}} f_\alpha\right)(x) = \begin{cases} \sup_{\alpha\in\mathbb{A}} f_\alpha(x)\ , & x \in D(\sup_{\alpha\in\mathbb{A}} f_\alpha) \\ \infty\ , & x \notin D(\sup_{\alpha\in\mathbb{A}} f_\alpha), \end{cases}$$

where $D(\sup_{\alpha\in\mathbb{A}} f_\alpha) := \{x \in \mathscr{A}_+;\ \sup_{\alpha\in\mathbb{A}} f_\alpha(x) < \infty\}$, $\sup_{\alpha\in\mathscr{A}} f_\alpha$ is a weight in $\mathscr{A}$.

(2) Let $\{f_\lambda\}_{\lambda\in\Lambda}$ be a set of positive linear functionals on $\mathscr{A}$. Putting

$$\left(\sum_{\lambda\in\Lambda} f_\lambda\right)(x) = \begin{cases} \sum_{\lambda\in\Lambda} f_\lambda(x)\ , & x \in D(\sum_{\lambda\in\Lambda} f_\lambda) \\ \infty\ , & x \notin D(\sum_{\lambda\in\Lambda} f_\lambda), \end{cases}$$

where $D(\sum_{\lambda\in\Lambda} f_\lambda) := \{x \in \mathscr{A}_+;\ \sum_{\lambda\in\Lambda} f_\lambda(x) < \infty\}$, $\sum_{\lambda\in\Lambda} f_\lambda$ is a weight in $\mathscr{A}$.

(3) For any positive subcone $\mathcal{P}$ of $\mathscr{A}_+$ we put

$$\varphi(x) = \begin{cases} 0\ , & x \in \mathcal{P}, \\ \infty\ , & x \notin \mathcal{P}. \end{cases}$$

Then φ is a weight in $\mathscr{A}$ with $D(\varphi) = \mathcal{P}$.

We here define the following notions needed later.

Definition 5.4.3 Let φ be a weight in $\mathscr{A}$. A weight ψ is called φ-dominated if

$$\psi(x) \leqq \gamma\varphi(x), \quad x \in \mathscr{A}_+$$

for some positive constant $\gamma > 0$, and then write $\psi \leqq \gamma\varphi$. A positive linear functional f on $\mathscr{A}$ is called φ-majorized if it is φ-dominated and there exists a positive constant $\gamma > 0$ such that

$$|f(x)|^2 \leqq \gamma\varphi(x^*x) \quad \text{for all} \quad x \in \mathscr{A}_\varphi.$$

A weight ψ in $\mathscr{A}$ is called an extension of φ if

$$D(\varphi) \subset D(\psi) \quad \text{and} \quad \varphi(x) = \psi(x) \quad \text{for all } x \in D(\varphi),$$

and then write $\varphi \subset \psi$.

We now have the following

Lemma 5.4.4 *Let φ be a weight in $\mathscr{A}$. Then every φ-dominated positive linear functional f on $\mathscr{A}$ is φ-majorized and there exists an element K of $\mathfrak{A}_\varphi^\tau$ such that $\pi(K) \geqq 0$ and $f(x) = (\lambda_\varphi(x)|\lambda(K))$ for all $x \in \mathscr{A}_\varphi$.*

Proof Let f be a φ-dominated positive linear functional on the C^*-algebra $\mathscr{A}$. Then f is f^0-majorized by Example 5.2.13, so that there exist a positive constant γ such that

$$|f(x)|^2 \leqq \gamma\varphi(x^*x)$$

for all $x \in \mathscr{A}_\varphi$. Hence, f is a φ-majorized positive linear functional on $\mathscr{A}$, which implies by Lemma 5.2.10 that there exists an element K of $\mathfrak{A}_\varphi^\tau$ such that $\pi(K) \geqq 0$ and $f(x) = (\lambda_\varphi(x)|\lambda(K))$ for all $x \in \mathscr{A}_\varphi$. This completes the proof.

For the regularity of weights we get the following result. This is the main consequence of this section.

Theorem 5.4.5 *Let φ be a weight in a C^*-algebra $\mathscr{A}$. Then the following statements are equivalent:*

(i) φ is regular.
(ii) $\varphi \subset \sup_{\alpha\in\mathbb{A}} f_\alpha$ for some net $\{f_\alpha\}_{\alpha\in\mathbb{A}}$ of positive linear functionals on $\mathscr{A}$.
(iii) $\varphi \subset \sum_{\lambda\in\Lambda} f_\lambda$ for some set $\{f_\lambda\}_{\lambda\in\Lambda}$ of positive linear functionals on $\mathscr{A}$.

Proof (i)⇒(iii) Since $\varphi = \varphi_r$, we have

$$\|F_{\mathfrak{A}_\varphi}\lambda_\varphi(x)\| = \|\lambda_\varphi(x)\| \quad \text{for all} \quad x \in \mathscr{A}_\varphi,$$

which implies that $\mathfrak{A}_\varphi$ is semisimple. Indeed, for an arbitrary $A \in N(\mathfrak{A}_\varphi)$, we can take a sequence $\{x_n\}$ in $\mathscr{A}_\varphi$ such that $\pi_\varphi(x_n) \to 0$ uniformly, $\lim_{n\to\infty} \lambda_\varphi(x_n) = \lambda(A)$ and $\lim_{n\to\infty} \lambda_\varphi(x_n^*) = \lambda(A^\sharp)$. Since $F_{\mathfrak{A}_\varphi}\lambda(A) = P_{\mathfrak{A}_\varphi^\tau} P_{\mathfrak{A}_\varphi}\lambda(A) = P_{\mathfrak{A}_\varphi^\tau}\lambda(A)$, it follows that

$$
\begin{aligned}
\|\lambda(A)\| &= \lim_{n\to\infty} \|\lambda_\varphi(x_n)\| \\
&= \lim_{n\to\infty} \|F_{\mathfrak{A}_\varphi}\lambda_\varphi(x_n)\| \\
&= \|F_{\mathfrak{A}_\varphi}\lambda(A)\| \\
&= \|P_{\mathfrak{A}_\varphi^\tau}\lambda(A)\| \\
&= \lim_\alpha \|\pi(e'_\gamma)\lambda(A)\| \\
&= \lim_\alpha \|\pi(A)\lambda(e'_\gamma)\| \\
&= 0,
\end{aligned}
$$

where $\{e'_\gamma\}$ is a net in $\mathfrak{A}_\varphi^\tau$ such that $\pi(e'_\gamma) \to P_{\mathfrak{A}_\varphi^\tau}$ strongly. Similarly, $\|\lambda(A^\sharp)\| = 0$. Thus, $A = O$. Therefore, $N_{\mathfrak{A}_\varphi} = \{O\}$ and $\mathfrak{A}_\varphi$ is semisimple. By Theorem 4.2.6 there exists an element $\{g_\lambda\}_{\lambda\in\Lambda}$ of $\mathfrak{M}_{\mathfrak{A}_\varphi}$ such that

$$(\pi_\varphi(x)g_\lambda|g_\mu) = 0, \quad \lambda \neq \mu$$

and

$$\lambda_\varphi(x) = \sum_{\lambda\in\Lambda} \pi_\varphi(x)g_\lambda$$

for all $x \in \mathscr{A}_\varphi$. Here, for any $\lambda \in \Lambda$ we put

$$f_\lambda(a) = (\pi_\varphi(a)g_\lambda|g_\lambda), \quad a \in \mathscr{A}.$$

Then f_λ is a positive linear functional on $\mathscr{A}$ and $\varphi(x) = \sum_{\lambda\in\Lambda} f_\lambda(x)$ for all $x \in D(\varphi)$. Hence, $\varphi \subset \sum_{\lambda\in\Lambda} f_\lambda$.

(iii)$\Rightarrow$(ii) This is trivial.

(ii)$\Rightarrow$(i) Take an arbitrary $a \in \mathscr{A}_\varphi$. Since $\sup_{\alpha\in\mathbb{A}} f_\alpha(a^*a) = \varphi(a^*a)$, for any $\varepsilon > 0$ we can choose an element α of $\mathbb{A}$ such that

$$f_\alpha(a^*a) + \varepsilon > \varphi(a^*a). \tag{5.4.1}$$

Since $f_\alpha \leqq \varphi$, it follows from Lemma 5.4.4 that there exists an element K of $\mathfrak{A}_\varphi^\tau$ such that $0 \leqq \pi(K) \leqq I$ and $f_\alpha(x) = (\lambda_\varphi(x)|\lambda(K))$ for all $x \in \mathscr{A}_\varphi$, which implies by (5.4.1) that

$$\begin{aligned}(F_{\mathfrak{A}_\varphi}\lambda_\varphi(a)|\lambda_\varphi(a)) + \varepsilon &= (P_{\mathfrak{A}_\varphi^\tau}\lambda_\varphi(a)|\lambda_\varphi(a)) + \varepsilon \\ &\geqq (\pi(K)P_{\mathfrak{A}_\varphi^\tau}\lambda_\varphi(a)|\lambda_\varphi(a)) + \varepsilon \\ &= (\pi(K)\lambda_\varphi(a)|\lambda_\varphi(a)) + \varepsilon \\ &= f_\alpha(a^*a) + \varepsilon \\ &> \varphi(a^*a) \\ &= \|\lambda_\varphi(a)\|^2.\end{aligned}$$

Hence we have

$$\|(I - F_{\mathfrak{A}_\varphi})\lambda_\varphi(a)\| < \varepsilon^{\frac{1}{2}},$$

and $F_{\mathfrak{A}_\varphi}\lambda_\varphi(a) = \lambda_\varphi(a)$ as $\varepsilon \to 0$. Thus, $\varphi = \varphi_r$. This completes the proof.

For the singularity of weights in $\mathscr{A}$ we have the following

Theorem 5.4.6 *Let φ be a weight in a C^*-algebra $\mathscr{A}$. Then the following statements are equivalent:*

(i) φ is singular.
(ii) There does not exists any φ-dominated positive linear functional f on $\mathscr{A}$ such that $f\lceil_{D(\varphi)} \neq 0$.

Proof (i)⇒(ii) Since $\varphi = \varphi_s$, we have

$$P_{\mathfrak{A}_\varphi^\tau}\lambda_\varphi(x) = 0 \quad \text{for all} \quad x \in \mathscr{A}_\varphi. \tag{5.4.2}$$

Let f be any φ-dominated positive linear functional on $\mathscr{A}$. By Lemma 5.4.4 there exists an element K of $\mathfrak{A}_\varphi^\tau$ such that $\pi(K) \geqq 0$ and

$$f(x) = (\lambda_\varphi(x)|\lambda(K)) \quad \text{for all} \quad x \in \mathscr{A}_\varphi.$$

Then it follows from (5.4.2) that

$$\begin{aligned}f(x^*x) &= (\pi(K)\lambda_\varphi(x)|\lambda_\varphi(x)) \\ &= (P_{\mathfrak{A}_\varphi^\tau}\pi(K)\lambda_\varphi(x)|\lambda_\varphi(x)) \\ &= 0\end{aligned}$$

for all $x \in \mathscr{A}_\varphi$, which implies (ii).

(ii)⇒(i) Suppose now that there exists an element x_0 of $\mathscr{A}_\varphi$ such that $P_{\mathfrak{A}_\varphi^\tau}\lambda_\varphi(x_0) \neq 0$. Then,

$$\pi(K)\lambda_\varphi(x_0) \neq 0 \tag{5.4.3}$$

for some $K \in \mathfrak{A}_\varphi^\tau$ with $\pi(K) \geqq 0$. We here put

$$f(a) = (\pi_\varphi(a)\lambda(K)|\lambda(K)), \quad a \in \mathscr{A}.$$

Since

$$\begin{aligned} f(x^*x) &= \|\pi_\varphi(x)\lambda(K)\|^2 \\ &= \|\pi(K)\lambda_\varphi(x)\|^2 \\ &\leqq \|\pi(K)\|^2\varphi(x^*x) \end{aligned}$$

for all $x \in \mathscr{A}_\varphi$, it follows that f is a φ-dominated positive linear functional on $\mathscr{A}_\varphi$, and furthermore, by (5.4.3), $f(x_0^*x_0) = \|\pi(K)\lambda_\varphi(x_0)\|^2 \neq 0$. This is a contradiction. Thus, (ii)⇒(i) holds. This completes the proof.

Notes Such a study of weights in C^*-algebras as in this section has been noted in [[42] Section 4]. In Sect. 4.4.5 we have defined a regular weight in the von Neumann algebra $\pi(\mathfrak{A})''$ from a T^*-algebra $\mathfrak{A}$. Weights in operator algebras have been studied by G.K. Pederson [26], F. Combes [6], U. Haagerup [15] and others.

Appendix A
Functional Calculus, Polar Decomposition and Spectral Resolution for Bounded Operators on a Hilbert Space

Let $\mathcal{H}$ be a Hilbert space and write $\mathcal{H}_1 = \{x \in \mathcal{H};\ \|x\| = 1\}$.

A.1 Spectrum

The spectrum $S_p(A)$ of $A \in B(\mathcal{H})$ is defined by

$S_p(A) = \{\lambda \in \mathbb{C};\quad (\lambda I - A)$ dose not have the bounded inverse in $B(\mathcal{H})\}$.

Then $S_p(A)$ is a non-empty compact subset of $\mathbb{C}$, and the spectrum radius $r(A) := \sup_{\lambda \in S_p(A)} |\lambda|$ equals $\lim_{n\to\infty} \|A^n\|^{\frac{1}{n}}$. In particular, if A is self-adjoint, then

$$S_p(A) \subset [m_A, M_A] \quad \text{and} \quad m_A, M_A \in S_p(A),$$

where

$$m_A := \inf\{(Ax|x);\quad x \in \mathcal{H}_1\},$$
$$M_A := \sup\{(Ax|x);\quad x \in \mathcal{H}_1\}.$$

A.2 Continuous Functional Calculus of a Self-Adjoint Bounded Linear Operator

Let $A = A^* \in B(\mathcal{H})$. We denote by $C(S_p(A))$ the C^*-algebra of all complex-valued continuous functions on the compact space $S_p(A)$ equipped with the usual function operators $f + g, \alpha f$ and fg, the involution $f \to f^* = \bar{f}$ and the uniform

A. Inoue, *Tomita's Lectures on Observable Algebras in Hilbert Space*,
Lecture Notes in Mathematics 2285, https://doi.org/10.1007/978-3-030-68893-6

norm $\|f\|_u := \sup_{t\in S_p(A)} |f(t)|$, and denote by $C^*(A, I)$ the C^*-algebra generated by $\{A, I\}$. Then we have the following

Theorem A.1 (Continuous Functional Calculus) *Let $A = A^* \in B(\mathcal{H})$. Then there exists a unique mapping:*

$$f \in C(S_p(A)) \longrightarrow f(A) \in C^*(A, I)$$

such that

(i) if f is a polynomial, $f(t) = \alpha_0 + \alpha_1 t + \cdots + \alpha_n t^n$, then $f(A) = \alpha_0 I + \alpha_1 A + \cdots + \alpha_n A^n$,
(ii) $\|f\|_u = \|f(A)\|$ for all $f \in C(S_p(A))$.

Furthermore, this mapping is a $*$-isomorphism of the C^*-algebra $C(S_p(A))$ onto the C^*-algebra $C^*(A, I)$.

A.3 Polar Decomposition for a Bounded Linear Operator

A bounded linear operator A on $\mathcal{H}$ is called *positive* if $S_p(A) \subset \mathbb{R}_+ := \{t \in \mathbb{R};\ t \geqq 0\}$, and then denote by $A \geqq 0$. We denote by $B(\mathcal{H})_+$ the set of all positive bounded linear operators on $\mathcal{H}$, and then it is a positive cone, that is, $A + B, \alpha A \in B(\mathcal{H})_+$ whenever $A, B \in B(\mathcal{H})_+$ and $\alpha \geqq 0$ and $B(\mathcal{H})_+ \cap (-B(\mathcal{H})_+) = \{0\}$. Using the functional calculus theorem (Theorem A.1), $A \in B(\mathcal{H})_+$ if and only if $A = B^2$ for the unique $B \in B(\mathcal{H})_+$. This B is called the square root of A and denoted by $A^{\frac{1}{2}}$ and $A^{\frac{1}{2}} \in C^*(A, I)$. An element U of $B(\mathcal{H})$ is a *partial isometry* if there exists a closed subspace $\mathcal{K}$ of $\mathcal{H}$ such that

$$\|Ux\| = \|x\| \quad \text{for all} \quad x \in \mathcal{K}$$

and

$$Uy = 0 \quad \text{for all} \quad y \in \mathcal{K}^\perp,$$

and $\mathcal{K}$ is called the *initial subspace* and $U\mathcal{K}$ is called the *final subspace* of U. Let U be a partial isometry with the initial subspace $\mathcal{K}$. Then U^*U and UU^* are projections such that $U^*U\mathcal{H} = \mathcal{K}$ and $UU^*\mathcal{H} = U\mathcal{K}$, and U^*U and UU^* are called the *initial projection* and the *final projection* of U, respectively.

Theorem A.2 (Polar Decomposition) *Let $A \in B(\mathcal{H})$. Then there exists a unique pair (U, B) of a partial isometry and an element B of $B(\mathcal{H})_+$ such that*

$$A = UB \quad \text{and} \quad U^*U\mathcal{H} = [B\mathcal{H}].$$

Here, $B = (A^*A)^{\frac{1}{2}}$ and it is called the absolute value of A and is denoted by $|A|$. The relation $A = U|A|$ is called the polar decomposition of A.

Let $A = U|A|$ be the polar decomposition of $A \in B(\mathcal{H})$. Then, $A^* = U^*(U|A|U^*) = U^*|A^*|$ is the polar decomposition of A^*, and in particular, if $A = A^*$, then $U^* = U$.

A.4 Spectral Resolution for a Bounded Self-Adjoint Operator

A family $\{E(\lambda);\ \lambda \in \mathbb{R}\}$ of projections on $\mathcal{H}$ is called a *spectral family* or a *resolution of the identity* if

(i) $E(\mu) \leqq E(\lambda)$ whenever $\mu \leqq \lambda$;
(ii) $E(\lambda + 0) := \lim_{\lambda_n \downarrow \lambda} E(\lambda_n)$, strongly $= E(\lambda)$, $\lambda \in \mathbb{R}$;
(iii) $E(\infty) := \lim_{\lambda_n \uparrow \infty} E(\lambda_n)$, strongly $= I$ and $E(-\infty) := \lim_{\lambda_n \downarrow -\infty} E(\lambda_n)$, strongly 0.

Let $A = A^* \in B(\mathcal{H})$. For any $\lambda \in \mathbb{R}$ we define a sequence $\{f_n^\lambda\}$ of continuous functions on $\mathbb{R}$ by

$$f_n^\lambda(t) = \begin{cases} 1 & , \quad -\infty < t \leqq \lambda \\ -n(t-\lambda)+1 & , \quad \lambda \leqq t \leqq \lambda + \frac{1}{n} \\ 0 & , \quad \lambda + \frac{1}{n} \leqq t < \infty, \end{cases}$$

and get

$$f_n^\lambda(t) \downarrow \chi_{(-\infty,\lambda]}(t) := \begin{cases} 1, & t \in (-\infty, \lambda] \\ 0 & , t \in (\lambda, \infty) \end{cases}$$

$$(f_n^\lambda)^2(t) \downarrow \chi_{(-\infty,\lambda]}(t) \quad , \quad t \in \mathbb{R}.$$

Then we can prove that there exists a projection $E_A(\lambda)$ on $\mathcal{H}$ such that $f_n^\lambda(A) \downarrow E_A(\lambda)$, strongly and $\{E_A(\lambda);\ \lambda \in \mathbb{R}\}$ is a spectral family in the von Neumann algebra $\{A\}''$ generated by A. For any $\delta > 0$ we consider a partition

$$\Delta = \{\lambda_0 = m_A - \delta < \lambda_1 < \cdots < \lambda_n = M_A\}$$

and put

$$s_\Delta = \sum_{k=1}^{n} \lambda_{k-1}(E_A(\lambda_k) - E_A(\lambda_{k-1})),$$

$$S_\Delta = \sum_{k=1}^{n} \lambda_k(E_A(\lambda_k) - E_A(\lambda_{k-1})).$$

Then, $\lim_{\|\triangle\|\to 0} S_\triangle = \lim_{\|\triangle\|\to 0} s_\triangle = A$ uniformly, where $\| \triangle \| := \max_{1\leqq k\leqq n} |\lambda_k - \lambda_{k-1}|$ and these limits are independent of the method of taking the partition $\triangle$ and $\delta > 0$. We here denote, as the usual integral, $\lim_{\triangle} S_\triangle = \lim_{\triangle} s_\triangle$ by $\int_{m_A-0}^{M_A} \lambda dE_A(\lambda)$. Thus, we have the following

Theorem A.3 (Spectral Resolution) *Let $A = A^* \in B(\mathcal{H})$. Then there exists a unique spectral family $\{E_A(\lambda);\ \lambda \in \mathbb{R}\}$ in $\{A\}''$ such that*

$$A = \int_{m_A-0}^{M_A} \lambda dE_A(\lambda), \quad E_A(m_A - 0) = 0, \quad E_A(M_A) = I.$$

This is called the spectral resolution of A.

A.5 Functional Calculus with Borel Functions

Let $A = A^* \in B(\mathcal{H})$. For any $f \in C(S_p(A))$ the norm-convergent vector stieltjes integral

$$\int_{m_A-0}^{M_A} f(\lambda)dE_A(\lambda)$$

is similarly defined and equals $f(A)$ given in Theorem A.1. We extend Theorem A.1 of the functional calculus with continuous functions to that with Borel functions. We denote by $B(S_p(A))$ the C^*-algebra of all bounded complex Borel functions on $S_p(A)$ with usual operations and involution and with the uniform norm $\| \cdot \|_u$. Let $\{E_A(\lambda);\ \lambda \in \mathbb{R}\}$ be a spectral family in $\{A\}''$. For any $x, y \in \mathcal{H}$ we put

$$\mu_{x,y}(\lambda) := (E_A(\lambda)x|y), \quad \lambda \in \mathbb{R}.$$

Then $\mu_{x,y}$ is a function of bounded variation and the total variation $V(\mu_{x,y})$ is majorized by $\|x\|\|y\|$, and so $\mu_{x,y}$ determines a bounded Borel measure whose support is contained in $S_p(A)$. For any $f \in B(S_p(A))$ we can define a bounded sesquilinear form F_f on $\mathcal{H} \times \mathcal{H}$ by the Lebesgue Stieltjes integral as follows:

$$F_f(x, y) = \int_{m_A-0}^{M_A} f(\lambda)d\mu_{x,y}(\lambda), \quad x, y \in \mathcal{H}.$$

By the Riesz theorem there exists a unique bounded linear operator $f(A)$ on $\mathcal{H}$ such that

$$(f(A)x|y) = \int_{m_A-0}^{M_A} f(\lambda)d\mu_{x,y}(\lambda), \quad x, y \in \mathcal{H}.$$

We now have the following

Theorem A.4 (Functional Calculus with Borel Functions) *Let $A = A^* \in B(\mathcal{H})$. Then the map: $f \to f(A)$ is a $*$-homomorphism of the C^*-algebra $B(S_p(A))$ into the von Neumann algebra $\{A\}''$ such that*

(i) if $f(\lambda) = \alpha_0 + \alpha_1\lambda + \cdots + \alpha_n\lambda^n$, $\lambda \in S_p(A)$, then $f(A) = \alpha_0 I + \alpha_1 A + \cdots + \alpha_n A^n$;

(ii) if $\{f_n\}$ is a sequence in $B(S_p(A))$ which converges pointwise to an element f of $B(S_p(A))$ such that $\sup_n \|f_n\|_u < \infty$, then $\lim_{n\to\infty} f_n(A) = f(A)$, strongly;

(iii) it is an extension of the $*$-isomorphism given by Theorem A.1.

Furthermore, the map: $f \in B(S_p(A)) \to f(A) \in B(\mathcal{H})$ satisfying (i) and (ii) is unique.

We refer to [29, 35] for Appendix A.

Appendix B
Spectral Resolution of an Unbounded Self-Adjoint Operator and Polor Decomposition of a Closed Operator in a Hilbert Space

B.1 Basic Definitions and Results for Unbounded Linear Operators

A linear operator T in a Hilbert space $\mathcal{H}$ is a linear mapping from the subspace $D(T)$ (called the *domain* of T) into $\mathcal{H}$. Let S and T be linear operators in $\mathcal{H}$. If $D(S) = D(T)$ and $Sx = Tx$ for all $x \in D(S) = D(T)$, then S and T are *equal* and write $S = T$. If $D(S) \subset D(T)$ and $Sx = Tx$ for all $x \in D(S)$, then T is an *extension* of S and denote by $S \subset T$. The algebraic operations of S, T and $\alpha \in \mathbb{C}$ are defined as follows:

(i) Addition $S + T$:

$$\begin{cases} D(S+T) = D(S) \cap D(T), \\ (S+T)x = \quad Sx + Tx, \quad x \in D(S+T); \end{cases}$$

(ii) The multiplication αT by scalar $\alpha \in \mathbb{C}$: If $\alpha = 0$, then $\alpha T = 0$, otherwise

$$\begin{cases} D(\alpha T) \ = D(T), \\ (\alpha T)(x) = \alpha(Tx), \ x \in D(T); \end{cases}$$

(iii) The multiplication ST:

$$D(ST) = \{x \in D(T); \quad Tx \in D(S)\},$$

$$(ST)x = S(Tx), \quad x \in D(ST).$$

A. Inoue, *Tomita's Lectures on Observable Algebras in Hilbert Space*,
Lecture Notes in Mathematics 2285, https://doi.org/10.1007/978-3-030-68893-6

If T is injective, then the inverse T^{-1} of T is defined by

$$\begin{cases} D(T^{-1}) = R(T), \\ T^{-1}(Tx) = \quad x, \quad x \in D(T). \end{cases}$$

The usual associative laws hold for the addition and for the multiplication: $(A + B) + C = A + (B + C)$ and $(AB)C = A(BC)$. For the distribute laws $(A + B)C = AC + BC$ and $A(B + C) \supset AB + AC$ hold, but $A(B + C) \subset AB + AC$ does not necessarily hold. Let T be a linear operator in $\mathcal{H}$. The set

$$G(T) := \{(x, Tx); \quad x \in D(T)\}$$

is a subspace of the direct sum $\mathcal{H} \oplus \mathcal{H}$, called the *graph* of T. It is clear that $S = T$ if and only if $G(S) = G(T)$, and $S \subset T$ if and only if $G(S) \subset G(T)$. A linear operator T in $\mathcal{H}$ is called *closed* if the graph $G(T)$ is closed in $\mathcal{H} \oplus \mathcal{H}$, that is, if $\{x_n\} \subset D(T)$ such that $x_n \to x$ and $Tx_n \to y$, then $x \in D(T)$ and $y = Tx$. Let T be a closed linear operator in $\mathcal{H}$. Then $D(T)$ is a Hilbert space with respect to the inner product:

$$(x|y)_T = (x|y) + (Tx|Ty), \quad x, y \in D(T).$$

This Hilbert space is denoted by $\mathcal{H}_T$, and the norm $\|\cdot\|_T$ defined by the inner product $(\cdot|\cdot)_T$ is called the *graph norm* associated to T. A linear operator T in $\mathcal{H}$ is called *closable* (or *pre-closed*) if it has a closed extension. For characterization of closable linear operators we have the following

Proposition B.1 *Let T be a linear operator in $\mathcal{H}$. Then the following statements are equivalent:*

(i) T is closable.
(ii) There exists a closed linear operator S in $\mathcal{H}$ such that $G(S) = \overline{G(T)}$.
(iii) If $\{x_n\} \subset D(T)$ such that $x_n \to 0$ and $Tx_n \to y$, then $y = 0$.

In this case, S in (ii) is the smallest closed extension of T, called the closure of T and denoted by $\bar{T}$, and $\bar{T}$ is given by

$$\begin{cases} D(\bar{T}) = \{x \in \mathcal{H} \text{ ; there exists a sequence } \{x_n\} \text{ in } D(T) \\ \qquad\qquad \text{such that } x_n \to x \text{ and } Tx_n \to y \in \mathcal{H}\}, \\ \bar{T}x \;=\; y, \qquad x \in D(\bar{T}). \end{cases}$$

Next we define the *adjoint* T^* of a densely defined linear operator T in $\mathcal{H}$ as follows:

$$\begin{cases} D(T^*) = \{y \in \mathcal{H} \text{ ; there exists an element } z \text{ of } \mathcal{H} \text{ such} \\ \qquad\qquad \text{that } (Tx|y) = (x|z) \text{ for all } x \in D(T)\}, \\ T^*y \;=\; z, \qquad y \in D(T^*). \end{cases}$$

Since $D(T)$ is dense in $\mathcal{H}$, T^* is a well-defined linear operator in $\mathcal{H}$, but $D(T^*)$ is not necessarily dense in $\mathcal{H}$. For adjoint operators we have the following

Proposition B.2 *Let S and T be densely defined linear operators in $\mathcal{H}$. Then the following statements hold:*

(1) If $S \subset T$, then $T^* \subset S^*$.
(2) $(\alpha T)^* = \bar{\alpha} T^*$, $\alpha \in \mathbb{C}$.
(3) If $D(S+T)$ is dense in $\mathcal{H}$, then $(S+T)^* \supset S^* + T^*$. In particular, if S or $T \in B(\mathcal{H})$, then $(S+T)^* = S^* + T^*$.
(4) If $D(ST)$ is dense in $\mathcal{H}$, then $(ST)^* \supset T^* S^*$. In particular, if $S \in B(\mathcal{H})$, then $(ST)^* = T^* S^*$.

Proposition B.3 *Let T be a densely defined linear operator in $\mathcal{H}$. Then the following statements hold:*

(1) T^* is closed.
(2) T is closable if and only if $D(T^*)$ is dense in $\mathcal{H}$. In this case $\bar{T} = T^{**} := (T^*)^*$.
(3) If T is closable, then $(\bar{T})^* = T^*$.

We define the notions of symmetric and self-adjoint operators: A densely defined linear operator T in $\mathcal{H}$ is called *symmetric* if $T \subset T^*$, and T is called *self-adjoint* if $T = T^*$. The distinction between closed symmetric operators and self-adjoint operators is very important. We discuss now the self-adjointness of closed symmetric operators. Let T be a symmetric operator in $\mathcal{H}$. We put

$$D_+(T) = \ker(T^* - iI) = \{y \in D(T^*);\ T^* y = iy\},$$
$$D_-(T) = \ker(T^* + iI) = \{y \in D(T^*);\ T^* y = -iy\}.$$

Then, $D_\pm(T)$ are closed subspaces of $\mathcal{H}$ satisfying $D_\pm(T) = R(T \pm iI)^\perp$. We here put

$$U_T(T + iI)x = (T - iI)x, \quad x \in D(T).$$

Then it is an isometry of $R(T+iI)$ onto $R(T-iI)$, so that it can be extended to the isometry of $[R(T+iI)]$ onto $[R(T-iI)]$ and denoted by the same U_T. This U_T is called the *Cayley tramsform* of T. Furthermore, T is closed if and only if $R(T \pm iI)$ are closed subspaces of $\mathcal{H}$. For the self-adjointness of closed symmetric operators we have the following

Proposition B.4 *Let T be a closed symmetric operator in $\mathcal{H}$. Then the following statements are equivalent:*

(i) T is self-adjoint.
(ii) $D_\pm(T) = \{0\}$.
(iii) $R(T \pm iI) = \mathcal{H}$.
(iv) The Cayley transform U_T of T is unitary.

B.2 Spectral Resolution of Unbounded Self-Adjoint Operators

We extend to unbounded self-adjoint operators the spectral theorem for bounded self-adjoint operators. A linear operator T in $\mathcal{H}$ *commutes* with a bounded linear operator A on $\mathcal{H}$ if $AT \subset TA$. Let $\mathscr{A}$ be a von Neumann algebra on $\mathcal{H}$. If T commutes with all operators in $\mathscr{A}'$, then we say that T is *affiliated* with $\mathscr{A}$.

Theorem B.5 (Spectral Resolution of Self-Adjoint Operators) *Let H be a self-adjoint operator in $\mathcal{H}$. Then its spectrum*

$$\begin{aligned} S_p(H) &:= \{\lambda \in \mathbb{C};\quad (\lambda I - H)^{-1} \textit{ does not exist in } B(\mathcal{H})\} \\ &\subset \mathbb{R}, \end{aligned}$$

and there exists a unique spectral family $\{E_H(\lambda);\ \lambda \in \mathbb{R}\}$ such that

(i)

$$\begin{aligned} D(H) &= \left\{x \in \mathcal{H};\ \int_{-n}^{n} \lambda dE_H(\lambda)x \textit{ converges to an element of } \mathcal{H}\right\} \\ &= \left\{x \in \mathcal{H};\ \int_{-n}^{n} \lambda dE_H(\lambda)x \textit{ converges weakly to an element of } \mathcal{H}\right\} \\ &= \left\{x \in \mathcal{H};\ \int_{-\infty}^{\infty} |\lambda|^2 d\|E_H(\lambda)x\|^2 < \infty\right\}; \end{aligned}$$

(ii) $Hx = \int_{-\infty}^{\infty} \lambda dE_H(\lambda)x := \lim_{n\to\infty} \int_{-n}^{n} \lambda dE_H(\lambda)x,\ x \in D(H)$;
(iii) *H is affiliated with $\{E_H(\lambda);\ \lambda \in \mathbb{R}\}''$.*

B.3 Functional Calculus for Unbounded Self-Adjoint Operators

Let H be a self-adjoint operator in $\mathcal{H}$ and $\{E_H(\lambda);\ \lambda \in \mathbb{R}\}$ a spectral family of H. We denote by $B_{cb}(\mathbb{R})$ the $*$-algebra of all Borel functions on $\mathbb{R}$ which are bounded on each compact subset of $\mathbb{R}$. For any $f \in B_{cb}(\mathbb{R})$ we define an operator $f(H)$ by

$$\begin{cases} D(f(H)) = \left\{x \in \mathcal{H};\ \int_{-n}^{n} f(\lambda)dE_H(\lambda)x \text{ converges in } \mathcal{H}\right\} \\ f(H)x \ = \lim_{n\to\infty} \int_{-n}^{n} f(\lambda)dE_H(\lambda)x, \quad x \in D(f(H)). \end{cases}$$

Then we have the following

$$D(f(H)) = \left\{x \in \mathcal{H};\ \int_{-n}^{n} f(\lambda)dE_H(\lambda)x \text{ converges weakly to an element of } \mathcal{H}\right\}$$
$$= \left\{x \in \mathcal{H};\ \int_{-\infty}^{\infty} |f(\lambda)|^2 d\|E_H(\lambda)x\|^2 < \infty\right\}$$

and

$$\|f(H)x\|^2 = \int_{-\infty}^{\infty} |f(\lambda)|^2 d\|E(\lambda)x\|^2, \quad x \in D(f(H)).$$

We here denote $f(H)$ by $\int_{-\infty}^{\infty} f(\lambda)dE_H(\lambda)$.

Theorem B.6 (Functional Calculus for Unbounded Self-Adjoint Operators) *Let H be a self-adjoint operator in $\mathcal{H}$ and $\{E_H(\lambda);\ \lambda \in \mathbb{R}\}$ the spectral family of H. Then, for any $f \in B_{cb}(\mathbb{R})$ $f(H)$ is a densely defined closed linear operator in $\mathcal{H}$ having the following properties: For any $f, g \in B_{cb}(\mathbb{R})$ and $\alpha \in \mathbb{C}$,*

(i) if $f(\lambda) = \alpha_0 + \alpha_1\lambda + \cdots + \alpha_n\lambda^n$, then $f(H) = \alpha_0 I + \alpha_1 H + \cdots + \alpha_n H^n$;
(ii) $(\alpha f)(H) = \alpha f(H)$;
(iii) $(f+g)(H) \supset f(H) + g(H)$;
(iv) $(fg)(H) \supset f(H)g(H)$;
(v) $f(H)^* = \bar{f}(H)$;
(vi) $f(H)$ is affiliated with $\{E_H(\lambda);\ \lambda \in \mathbb{R}\}''$.

B.4 Polar Decomposition of Closed Linear Operators

A linear operator T in $\mathcal{H}$ is called *positive* if $(Tx|x) \geqq 0$ for all $x \in D(T)$. To define the *square root* of a positive self-adjoint operator in $\mathcal{H}$ we prepare the following

Proposition B.7 *Let T be a densely defined closed operator in $\mathcal{H}$. Then the following statements hold:*

(1) $(I + T^*T)^{-1}$ *exists and it is a bounded self-adjoint operator on* $\mathcal{H}$.
(2) T^*T *is a positive and self-adjoint operator in* $\mathcal{H}$.
(3) $D(T^*T)$ *is a core for* T, *that is,* $D(T^*T)$ *is dense in the Hilbert space* $\mathcal{H}_T$.

Proposition B.8 *Let H be a self-adjoint operator in $\mathcal{H}$. Then the following statements are equivalent:*

(i) H *is positive.*
(ii) $S_p(H) \subset [0, \infty)$.
(iii) $H = \int_0^\infty \lambda dE_H(\lambda)$.

By Propositions B.7, B.8, and Theorem B.6 we have the following

Theorem B.9 *Let H be a positive self-adjoint operator in $\mathcal{H}$. Then there exists a unique positive self-adjoint operator A in $\mathcal{H}$ such that $A^2 = H$. This operator A is called the square root of H and denoted by $H^{\frac{1}{2}}$.*

Thus, the positive self-adjoint operator $(T^*T)^{\frac{1}{2}}$ for any densely defined closed linear operator T in $\mathcal{H}$ can be defined and it is called the *absolute value* of T and denoted by $|T|$. We now have the following

Theorem B.10 (Polar Decomposition of Closed Operators) *Every densely defined closed operator T in $\mathcal{H}$ is uniquely decomposed into $T = U_T|T|$, where U_T is a partial isometry with initial subspace $[R(|T|)]$. Then, $[R(|T|)] = [R(T^*)]$.*

We refer to [3, 29] for Appendix B.

Appendix C
Banach $*$-Algebras, C^*-Algebras, von Neumann Algebras and Locally Convex $*$-Algebras

C.1 Banach $*$-Algebras

A complex vector space $\mathscr{A}$ is called an *algebra* if a map: $(a, b) \in \mathscr{A} \times \mathscr{A} \to ab \in \mathscr{A}$ is defined and satisfies the following conditions:

(i) $a(bc) = (ab)c$,
(ii) $(a + b)c = ac + bc$ and $a(b + c) = ab + ac$,
(iii) $\alpha(ab) = (\alpha a)b = a(\alpha b)$

for all $a, b, c \in \mathscr{A}$ and $\alpha \in \mathbb{C}$. An element e of an algebra $\mathscr{A}$ is called *identity* (or *unit*) if $ae = ea = a$ for all $a \in \mathscr{A}$. The identity of $\mathscr{A}$ is unique. An algebra $\mathscr{A}$ is called a *$*$-algebra* if there exists a conjugate linear map $a \in \mathscr{A} \to a^* \in \mathscr{A}$ such that $(ab)^* = b^*a^*$ and $(a^*)^* = a$ for all $a, b \in \mathscr{A}$. The map: $a \to a^*$ is called an *involution* of $\mathscr{A}$. If e is the identity of a $*$-algebra, then $e^* = e$. A *normed $*$-algebra* $\mathscr{A}$ is a $*$-algebra which is also a normed space with norm $\| \cdot \|$ such that $\|ab\| \leqq \|a\|\|b\|$ and $\|a^*\| = \|a\|$ for all $a, b \in \mathscr{A}$ and $\|e\| = 1$ if $e \in \mathscr{A}$. A Banach $*$-algebra $\mathscr{A}$ is a normed $*$-algebra with norm $\| \cdot \|$ and $\mathscr{A}[\| \cdot \|]$ is a Banach space. Let $\mathscr{A}$ be a Banach $*$-algebra without identity. Then, the direct sum $\mathbb{C} \oplus \mathscr{A}$ as a vector space $\mathscr{A}_e$ is a Banach $*$-algebra with identity $(1, 0)$ equipped with the following multiplication, the involution and the norm:

$$(\alpha, a)(\beta, b) = (\alpha\beta, \alpha b + \beta a + ab),$$
$$(\alpha, a)^* = (\bar{\alpha}, a^*),$$
$$\|(\alpha, a)\| = |\alpha| + \|a\|$$

for all $(\alpha, a), (\beta, b) \in \mathbb{C} \oplus \mathscr{A}$. The map: $a \in \mathscr{A} \to (0, a) \in \mathscr{A}_e$ is an isometric and $*$-isomorphism. Identifying each $a \in \mathscr{A}$ and $(o, a) \in \mathscr{A}_e$, we write $(\alpha, a) = \alpha e + a \in \mathscr{A}_e$. This Banach $*$-algebra $\mathscr{A}_e$ is called the *adjunction of an identity* to $\mathscr{A}$, or the *unitization* of $\mathscr{A}$. We often enlarge the algebra by adjunction of an identity

A. Inoue, *Tomita's Lectures on Observable Algebras in Hilbert Space*,
Lecture Notes in Mathematics 2285, https://doi.org/10.1007/978-3-030-68893-6

to a given algebra. But, this approach does not always work well. For example, a positive linear functional on a $*$-algebra $\mathscr{A}$ without identity can not be extended to a positive linear functional on $\mathscr{A}_e$ in general. In this respect, the notion of an approximate identity is very useful. An *approximate identity* of a Banach $*$-algebra is a net $\{e_\alpha\}$ in $\mathscr{A}$ with the properties:

$$\lim_\alpha \|e_\alpha a - a\| = 0 \quad \text{and} \quad \lim_\alpha \|a e_\alpha - a\| = 0$$

for any $a \in \mathscr{A}$. If $\{e_\alpha\}$ is bounded, then $\{e_\alpha\}$ is called a *bounded approximate identity*.

C.2 C^*-Algebras and von Neumann Algebras

A normed $*$-algebra (resp. Banach $*$-algebra) $\mathscr{A}$ with norm $\|\cdot\|$ is called a *C^*-normed algebra* (resp. *C^*-algebra*) if $\|a^*a\| = \|a\|^2$ for all $a \in \mathscr{A}$. The Banach $*$-algebra $\mathscr{A}_e$ obtained by the adjunction of an identity of a C^*-algebra $\mathscr{A}$ without identity is not a C^*-algebra, but it is known that every C^*-algebra has a bounded approximate identity $\{e_\alpha\}$ satisfying more properties:

$$\begin{aligned} &0 \leqq e_\alpha \leqq e_\beta \quad \text{if} \quad \alpha \leqq \beta, \\ &\|e_\alpha\| \leqq 1. \end{aligned}$$

Let $B(\mathcal{H})$ be the set of all bounded linear operators on $\mathcal{H}$. Then $B(\mathcal{H})$ is a C^*-algebra equipped with the usual operations $A + B$, αA, AB, the involution $A \to A^*$ (the adjoint of A) and the uniform norm. A closed $*$-subalgebra of $B(\mathcal{H})$ is a C^*-algebra and it is called a *C^*-algebra on $\mathcal{H}$*. The celebrated Gelfand-Naimark theorem states that every (abstract) C^*-algebra is isometrically isomorphic to a C^*-algebra on some Hilbert space. Now we proceed to the definition of von Neumann algebras. Let $\mathscr{A}$ be a $*$-preserving subset of $B(\mathcal{H})$, that is, $A^* \in \mathscr{A}$ for all $A \in \mathscr{A}$. The *commutant* $\mathscr{A}'$ of $\mathscr{A}$ is defined by

$$\mathscr{A}' = \{C \in B(\mathcal{H});\ AC = CA \quad \text{for} \quad \text{all } A \in \mathscr{A}\},$$

and then $\mathscr{A}'$ is a $*$-algebra on $\mathcal{H}$ with identity operator I. We here call a $*$-subalgebra of $B(\mathcal{H})$ a *$*$-algebra on $\mathcal{H}$*. A $*$-algebra on $\mathcal{H}$ is called a *von Neumann algebra on $\mathcal{H}$* if $\mathscr{A} = \mathscr{A}'' := (\mathscr{A}')'$.

C.3 Density Theorems

An equivalent definition of a von Neumann algebra can be given in terms of some locally convex topologies on $B(\mathcal{H})$ that we now define. The locally convex topology on $B(\mathcal{H})$ generated by the family $\{p_{x,y};\ x, y \in \mathcal{H}\}$ (resp. $\{p_x;\ x \in \mathcal{H}\}$, $\{p_x^*;\ x \in \mathcal{H}\}$, $\{p_{\{x_n\},\{y_n\}};\ \{x_n\}, \{y_n\} \in \mathcal{H}_{\mathbb{N}}\}$, $\{p_{\{x_n\}};\ \{x_n\} \in \mathcal{H}_{\mathbb{N}}\}$ and $\{p^*_{\{x_n\},\{y_n\}};\ \{x_n\}, \{y_n\} \in \mathcal{H}_{\mathbb{N}}\}$ of seminorms is called the *weak* (resp. *strong, strong**, σ*-weak*, σ*-strong* and σ*-strong**) topology, where

$$
\begin{aligned}
p_{x,y}(A) &= |(Ax|y)|, \\
p_x(A) &= \|Ax\|, \\
p_x^*(A) &= \|Ax\| + \|A^*x\|, \\
\mathcal{H}_{\mathbb{N}} &= \{\{x_n\} \subset \mathcal{H};\ \sum_{n=1}^{\infty} \|x_n\|^2 < \infty\}, \\
p_{\{x_n\},\{y_n\}}(A) &= \sum_{n=1}^{\infty} |(Ax_n|y_n)|, \\
p_{\{x_n\}}(A) &= \left(\sum_{n=1}^{\infty} \|Ax_n\|^2\right)^{\frac{1}{2}}, \\
p^*_{\{x_n\}} &= p_{\{x_n\}}(A) + p_{\{x_n\}}(A^*).
\end{aligned}
$$

The fundamental results of the theory of von Neumann algebras are the following

Theorem C.1 (von Neumann Density Theorem) *Let $\mathscr{A}$ be a $*$-algebra on $\mathcal{H}$ and $\bar{\mathscr{A}}[\tau]$ the closure of $\mathscr{A}$ under any one τ of the weak, strong, strong***, σ-weak, σ-strong and σ-strong** *topologies. Then the following statements hold:*

(1) $P_{\mathscr{A}} := \mathrm{Proj}\,[\mathscr{A}\mathcal{H}] \in \bar{\mathscr{A}}[\tau]$ and $P_{\mathscr{A}}A = AP_{\mathscr{A}} = A$ for all $A \in \bar{\mathscr{A}}[\tau]$, in other words, $P_{\mathscr{A}}$ is the identity element of $\bar{\mathscr{A}}[\tau]$.
(2) $\mathscr{A}'' = \{\alpha I + A;\ \alpha \in \mathbb{C},\ A \in \bar{\mathscr{A}}[\tau]\}$. In particular, if $\mathscr{A}$ is nondegenerate, that is, $P_{\mathscr{A}} = I$, then $\mathscr{A}'' = \bar{\mathscr{A}}[\tau]$; (see e.g. [37, Theorem 3.9]).

Let $\mathscr{A}$ be a $*$-algebra on $\mathcal{H}$. We denote by $\mathscr{A}_h$ (resp. $\mathscr{A}_+$) the set of all self-adjoint (resp. positive) elements of $\mathscr{A}$, denote by $u(\mathscr{A})$ (resp. $u(\mathscr{A}_h)$, $u(\mathscr{A}_+)$) the unit ball of $\mathscr{A}$ (resp. $\mathscr{A}_h$, $\mathscr{A}_+$).

Theorem C.2 (Kaplansky Density Theorem) *Let $\mathscr{A}$ be a von Neumann algebra on $\mathcal{H}$ and $\mathscr{B}$ a $*$-subalgebra of $\mathscr{A}$. Suppose that $\mathscr{B}$ is strongly dense in $\mathscr{A}$. Then $u(\mathscr{B})$ (resp. $u(\mathscr{B}_h)$, $u(\mathscr{B}_+)$) is strongly dense in $u(\mathscr{A})$ (resp. $u(\mathscr{A}_h)$, $u(\mathscr{A}_+)$); (see e.g. [35, Theorem 3.10]).*

Let $\mathscr{A}$ be a $*$-algebra on $\mathcal{H}$. Then, since we get $P_{\mathscr{A}} \in \mathscr{A}' \cap \mathscr{A}''$, it follows that $\mathscr{A}_{P_{\mathscr{A}}} := \{A\lceil_{P_{\mathscr{A}}\mathcal{H}};\ A \in \mathscr{A}\}$ is a nondegenerate $*$-algebra on $P_{\mathscr{A}}\mathcal{H}$, which yields by Theorem C.1 that $(\mathscr{A}_{P_{\mathscr{A}}})'' = (\bar{\mathscr{A}}[\tau])_{P_{\mathscr{A}}}$, where τ is any one of the weak, strong, strong*, σ-weak, σ-strong and σ-strong* topologies. Hence, by Theorem C.2 we have the following result used often in this note.

Corollary C.3 *Let $\mathscr{A}$ be a $*$-algebra on $\mathcal{H}$. Then there exists a net $\{e_\alpha\}$ in $u(\mathscr{A}_+)$ which converges strongly to $P_{\mathscr{A}}$.*

C.4 Locally Convex $*$-Algebras

A *locally convex algebra* is a locally convex space which is also an algebra whose the multiplication $(a, b) \in \mathscr{A} \times \mathscr{A} \to ab \in \mathscr{A}$ is separately continuous, that is, for any $a \in \mathscr{A}$ the maps: $b \in \mathscr{A} \to ab \in \mathscr{A}$ and $b \in \mathscr{A} \to ba \in \mathscr{A}$ are continuous. A *locally convex $*$-algebra* is a locally convex algebra which is a $*$-algebra with continuous involution. Let $\mathcal{H}$ be a Hilbert space. Then $B(\mathcal{H})$ is a locally convex $*$-algebra under any one of the weak, σ-weak, strong* and σ-strong* topologies, and it is a locally convex algebra under any one of the strong and σ-strong topologies, but not necessarily a locally convex $*$-algebra.

For Appendix C,1, 4, we refer to [12, 20, 25, 30], and for Appendix C.2, 3 refer to [8, 9, 18, 32, 35, 37].

References

1. G.R. Allan, A spectral theory for locally convex algebras. Proc. Lond. Math. Soc. **15**, 399–421 (1965)
2. G.R. Allan, On a class of locally convex algebras. Proc. Lond. Math. Soc. **17**, 91–114 (1967)
3. J.P. Antoine, A. Inoue, C. Trapani, *Partial $*$-Algebras and Their Operator Realizations*. Math. Appl., vol. 553 (Kluwer Academic, Dordrect, 2002)
4. S.J. Bhatt, Representability of positive functionals on abstract star algebras without identity with applications to locally convex $*$-algebras. Yokohama Math. J. **29**, 7–16 (1981)
5. S.J. Bhatt, A. Inoue, H. Ogi, Admissibility of weights on non-normed $*$-algebras. Trans. Am. Math. Soc. **351**, 4629–4656 (1999)
6. F. Combes, Poids sur une C^*-algèbre. J. Math. Pures Appl. **42**, 57–100 (1968)
7. A. Connes, Une classification des facteurs de type III. Ann. Scient. École Norm. Sup. 4-ieme Sér **6**, 133–252 (1973)
8. J. Dixmier, *Les C^*-algèbres et leurs représentations* (Gauthier-Villars, Paris, 1964/1968)
9. J. Dixmier, *Les algèbres d'opérateurs dans l'espace hilbertien (Algèbres de von Neumann)* (Gauthier-Villars, Paris, 1957/1969)
10. P.G. Dixon, Generalized B^*-algebras. Proc. Lond. Math. Soc. **21**, 693–715 (1970)
11. M. Fragoulopoulou, An introduction to the representation theory of topological $*$-algebras. Schriftenreihe des Math. Inst. der Univ. Münster **48**, 1–81 (1988)
12. M. Fragoupoulou, *Topological Algebras with Involution* (North-Holland, Amsterdam, 2005)
13. I.M. Gelfand, N.Y. Vilenkin, *Generalized Functions*. Applications of Harmonic Analysis, vol. IV (Academic Press, New York, 1964)
14. I.M. Gelfand, M.I. Graev, N.Y. Vilenkin, *Generalized Functions*. Integral Geometry and Representation Theory, vol. V (Academic Press, New York, 1966)
15. U. Haagerup, Normal weights on W^*-algebras. J. Funct. Anal. **19**, 302–317 (1975)
16. A. Inoue, On regularity of positive linear functionals. Japanese. J. Math. **9**, 247–275 (1983)
17. A. Inoue, *Tomita-Takesaki Theory in Algebras of Unbounded Operators* (Springer, Berlin, 1998)
18. R.V. Kadison, J.R. Ringrose, *Fundamentals of the Theory of Operator Algebras*, vol. I (Academic Press, New York, 1983)
19. E.A. Michael, *Lacally Multiplicatively-Convex Topological Algebras*. Memoirs of the American Mathematical Society, vol. 11 (American Mathematical Society, Providence, 1953)
20. M.A. Naimark, *Normed Algebras* (Wolters-Noordhoff Publisher, Gröningen, 1972)
21. Y. Nakagami, Tomita's spectral analysis in Krein spaces. Publ. Res. Inst. Math. Sci. **22**, 637–658 (1986)
22. Y. Nakagami, Spectral analysis in Krein spaces. Publ. Res. Inst. Math. Soc. **24**, 361–378 (1988)

A. Inoue, *Tomita's Lectures on Observable Algebras in Hilbert Space*,
Lecture Notes in Mathematics 2285, https://doi.org/10.1007/978-3-030-68893-6

23. Y. Nakagami, M. Tomita, Triangular matrix representation for self-adjoint operators in Krein spaces. Japanese J. Math. **14**, 165-202 (1988)
24. S. Ôta, On a representation of a C^*-algebra in a Lorentz algebra. Acta Sci. Math. **39**, 129–133 (1977)
25. T.W. Palmer, *Banach Algebras and General Theory of $*$-Algebras*, vol. I (Cambridge University Press, Cambridge, 1994)
26. G.K. Pedersen, Measure theory for C^*-algebras. Math. Scand. **19**, 131–145 (1966)
27. J.D. Powell, Representations of locally convex $*$-algebras. Proc. Am. Math. Soc. **44**, 341–346 (1974)
28. R.T. Powers, Self-adjoint algebras of unbounded operators. Commun. Math. Phys. **21**, 55–124 (1971)
29. M.C. Reed, B. Simon, *Methods of Modern Mathematical Physics*. Functional Analysis, vol. I (Academic Press, New York, 1980)
30. C.E. Richart, *General Theory of Banach Algebras* (R.E. Krieger Publishing Company, Huntigton, 1974)
31. A.P. Robertson, W. Robertson, *Topological Vector Spaces* (Cambridge University Press, Cambridge, 1973)
32. S. Sakai, *C^*-Algebras and W^*-Algebras* (Springer, Berlin, 1998)
33. H.H. Schaefer, *Topological Vector Spaces* (Springer, New York, 1971)
34. K. Schmüdgen, *Unbounded Operator Algebras and Representation Theory* (Birkhäuser-Verlag, Basel, 1990)
35. S. Strătilă, L. Zsidó, *Lectures on von Neumann Algebras* (Abacus Press, Tunbrige Wells, 1979)
36. M. Takesaki, *Tomita's Theory of Modular Hilbert Algebras and its Applications*. Springer Lecture Notes in Mathematics, vol. 128 (Springer, Berlin, 1979)
37. M. Takesaki, *Theory of Operator Algebras I* (Springer, Berlin, 1979)
38. M. Takesaki, *Theory of Operator Algebras II* (Springer, Heidelberg, 1983)
39. M. Tomita, *Standard Forms of von Neumann Algebras*. The Fifth Functional Analysis Symposium of Mathematical Society of Japan, Sendai (1967)
40. M. Tomita, Harmonic analysis on topological $*$-algebras (in Japanese). RIMS-Kokyuroku **77**, 99–117 (1969)
41. M. Tomita, Algebra of observables in Hilbert space (in Japanese). RIMS-Kokyuroku **104**, 82–99 (1970)
42. M. Tomita, Foundations of noncommutative fourier analysis, in *Japan-U.S. Seminor on C^*-Algebras and Applications to Physics, Kyoto* (1974)
43. M. Tomita, Operators and operator algebras in Krein spaces, I Spectral analysis in Pontrjagin spaces. RIMS-Kokyuroku **398**, 131–158 (1980)
44. A. Van Daele, A new approach to Tomita-Takesaki theory of generalized Hilbert algebras. J. Funct. Anal. **15**, 378–393 (1974)

Index

A. Inoue, *Tomita's Lectures on Observable Algebras in Hilbert Space*,
Lecture Notes in Mathematics 2285, https://doi.org/10.1007/978-3-030-68893-6

LECTURE NOTES IN MATHEMATICS

Springer

Editors in Chief: J.-M. Morel, B. Teissier;

Editorial Policy

1. Lecture Notes aim to report new developments in all areas of mathematics and their applications – quickly, informally and at a high level. Mathematical texts analysing new developments in modelling and numerical simulation are welcome.

   Manuscripts should be reasonably self-contained and rounded off. Thus they may, and often will, present not only results of the author but also related work by other people. They may be based on specialised lecture courses. Furthermore, the manuscripts should provide sufficient motivation, examples and applications. This clearly distinguishes Lecture Notes from journal articles or technical reports which normally are very concise. Articles intended for a journal but too long to be accepted by most journals, usually do not have this "lecture notes" character. For similar reasons it is unusual for doctoral theses to be accepted for the Lecture Notes series, though habilitation theses may be appropriate.

2. Besides monographs, multi-author manuscripts resulting from SUMMER SCHOOLS or similar INTENSIVE COURSES are welcome, provided their objective was held to present an active mathematical topic to an audience at the beginning or intermediate graduate level (a list of participants should be provided).

   The resulting manuscript should not be just a collection of course notes, but should require advance planning and coordination among the main lecturers. The subject matter should dictate the structure of the book. This structure should be motivated and explained in a scientific introduction, and the notation, references, index and formulation of results should be, if possible, unified by the editors. Each contribution should have an abstract and an introduction referring to the other contributions. In other words, more preparatory work must go into a multi-authored volume than simply assembling a disparate collection of papers, communicated at the event.

3. Manuscripts should be submitted either online at www.editorialmanager.com/lnm to Springer's mathematics editorial in Heidelberg, or electronically to one of the series editors. Authors should be aware that incomplete or insufficiently close-to-final manuscripts almost always result in longer refereeing times and nevertheless unclear referees' recommendations, making further refereeing of a final draft necessary. The strict minimum amount of material that will be considered should include a detailed outline describing the planned contents of each chapter, a bibliography and several sample chapters. Parallel submission of a manuscript to another publisher while under consideration for LNM is not acceptable and can lead to rejection.

4. In general, **monographs** will be sent out to at least 2 external referees for evaluation.

   A final decision to publish can be made only on the basis of the complete manuscript, however a refereeing process leading to a preliminary decision can be based on a pre-final or incomplete manuscript.

   Volume Editors of **multi-author works** are expected to arrange for the refereeing, to the usual scientific standards, of the individual contributions. If the resulting reports can be forwarded to the LNM Editorial Board, this is very helpful. If no reports are forwarded or if other questions remain unclear in respect of homogeneity etc, the series editors may wish to consult external referees for an overall evaluation of the volume.

5. Manuscripts should in general be submitted in English. Final manuscripts should contain at least 100 pages of mathematical text and should always include

   – a table of contents;
   – an informative introduction, with adequate motivation and perhaps some historical remarks: it should be accessible to a reader not intimately familiar with the topic treated;
   – a subject index: as a rule this is genuinely helpful for the reader.
   – For evaluation purposes, manuscripts should be submitted as pdf files.

6. Careful preparation of the manuscripts will help keep production time short besides ensuring satisfactory appearance of the finished book in print and online. After acceptance of the manuscript authors will be asked to prepare the final LaTeX source files (see LaTeX templates online: https://www.springer.com/gb/authors-editors/book-authors-editors/manuscriptpreparation/5636) plus the corresponding pdf- or zipped ps-file. The LaTeX source files are essential for producing the full-text online version of the book, see http://link.springer.com/bookseries/304 for the existing online volumes of LNM). The technical production of a Lecture Notes volume takes approximately 12 weeks. Additional instructions, if necessary, are available on request from lnm@springer.com.

7. Authors receive a total of 30 free copies of their volume and free access to their book on SpringerLink, but no royalties. They are entitled to a discount of 33.3 % on the price of Springer books purchased for their personal use, if ordering directly from Springer.

8. Commitment to publish is made by a *Publishing Agreement*; contributing authors of multiauthor books are requested to sign a *Consent to Publish form*. Springer-Verlag registers the copyright for each volume. Authors are free to reuse material contained in their LNM volumes in later publications: a brief written (or e-mail) request for formal permission is sufficient.

Addresses:
Professor Jean-Michel Morel, CMLA, École Normale Supérieure de Cachan, France
E-mail: moreljeanmichel@gmail.com

Professor Bernard Teissier, Equipe Géométrie et Dynamique,
Institut de Mathématiques de Jussieu – Paris Rive Gauche, Paris, France
E-mail: bernard.teissier@imj-prg.fr

Springer: Ute McCrory, Mathematics, Heidelberg, Germany,
E-mail: lnm@springer.com

Printed in the United States
By Bookmasters